No Metal No Magic

Element 34

Selenium, Presented By Selenice

From The Magical Elements of the Periodic Table Book Series

Selenice

Selenium

By Sybrina Durant with Illustrations by Pranavva et al.

Selenium, Presented By Selenice
From The Magical Elements of the Periodic Table Book Series

Story copyright 2026

ISBN: 978-1-942740-58-2

BISAC Codes:

JNF051070 JUVENILE NONFICTION / Science & Nature / Chemistry

JNF016000 JUVENILE NONFICTION / Curiosities & Wonders

JNF051080 JUVENILE NONFICTION / Science & Nature / Earth Sciences / General

Soft Cover Print ISBN 13—978-1-942740-58-2

Selenice Presents Selenium

This Element 34 book features the periodic table element, Selenium. It is presented by Selenice, an Alchemical Wizard who wields a magical elemental staff with powers based on its periodic table element.

Selenice is just one of the 118 elementals who will present all of the Magical Elements of the Periodic Table to readers who are curious about the wonders of the world.

Selenice introduces the very magical element, Selinium, in her book.

The Alchemical Wizards and their other techno-magical friends are the perfect group to introduce you to the elements in the Periodic Table. Hopefully, this Magical Elements of the periodic table book will spark an interest in the magical and real world properties of all the elements known today. You may be surprised at how prominently they feature in our every day lives.

Each page in this book contains terms that might not be completely familiar to the reader. Refer to the definitions in the back of the book to get a clear understanding of each meaning.

There is also a fun elemental themed Periodic Table at the back of the book. It features 118 elements presented by fanciful characters like unicorns, dragons, wizards, knights and goblins.. They want you to remember that if there's no metal...there's no magic or technology.

Remember, "No metal – No Magic. . .and No Technology".

It's Techo-Magical.

Note: Sybrina Publishing websites are Sybrina.com and MagicalPTElements.com. Follow sybrinapublishing on Instagram, Magical Elements of the Periodic Table on Facebook, @sybrinad on Pinterest, Sybrina_SPT on Twitter; and Sybrina Durant on LinkedIn.

Selenium is a Non-Metal

Selenium was discovered in 1817 by Swedish chemist Jöns Jacob Berzelius (along with Johan Gottlieb Gahn) in Stockholm, Sweden. It was identified as a red-brown sediment in the sulfuric acid production chambers at a chemical plant in Gripsholm.

Selenium exists in three primary allotropic forms, primarily distinguished by their color and structure: metallic gray (stable, hexagonal, semiconductor), red (amorphous powder or monoclinic crystals), and black (vitreous/glassy solid). Gray selenium is the most stable form, while red and black forms are insulators that can be converted between types through heating and cooling.

Selenium is a p-type semiconductor (specifically in its gray crystalline form) that conducts electricity better than an insulator but not as well as a conductor.

Solid elemental selenium is weakly attracted to external magnetic fields. However, its magnetic properties can vary based on allotropic form or, in the case of biogenic selenium nanomaterials, show strong paramagnetic behavior.

Selinium is neither ductile nor malleable. It is brittle and if subjected to force, it will break or turn into powder rather than changing shape.

Selenium (atomic number 34) is a Non-Metal element. Being a non-metal means Selenium is more of a "taker" or "sharer" of electrons rather than a "giver" (like metals are).

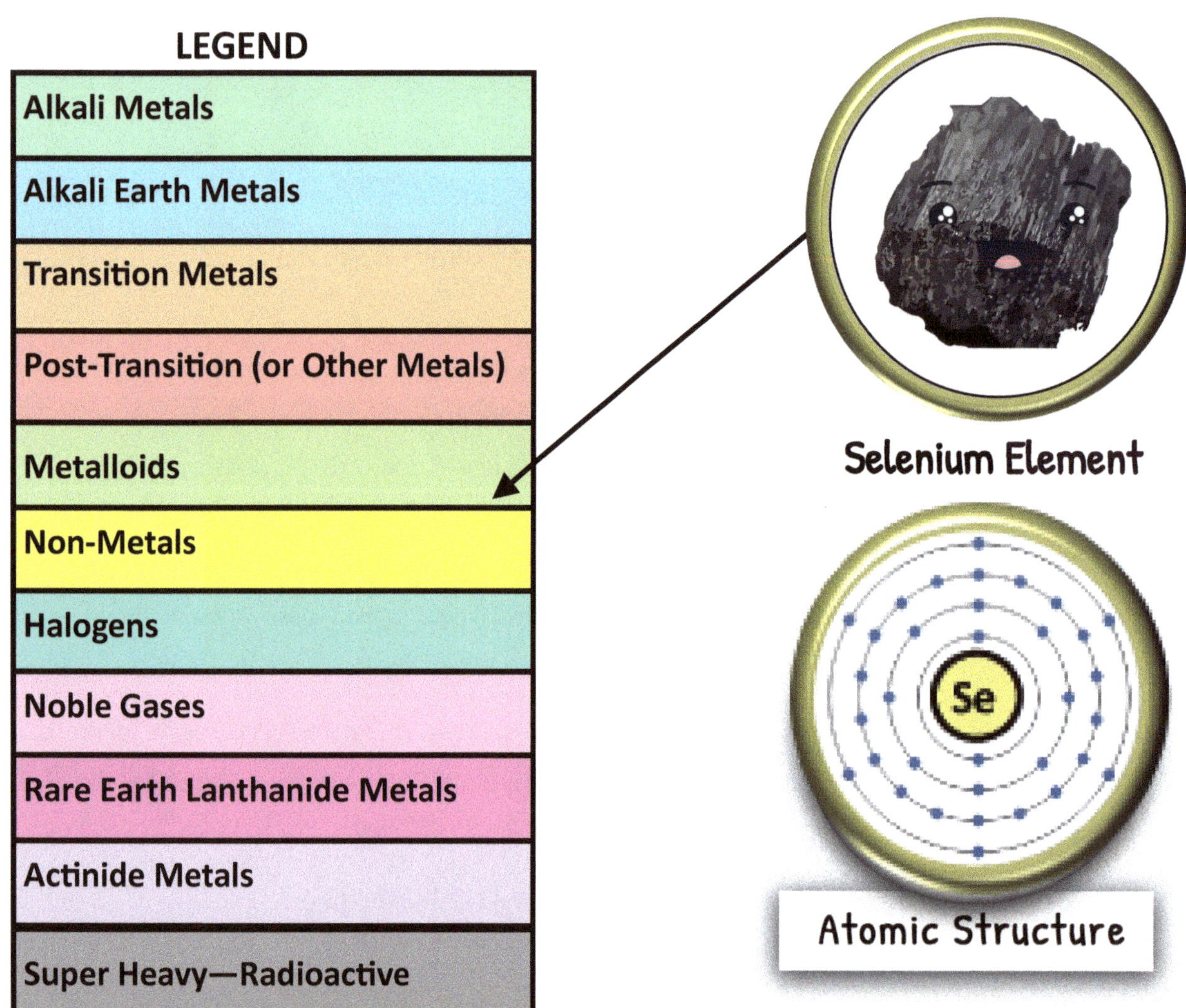

Selenium Element

Atomic Structure

Non-Metals—These elements reside in columns 15-17, and can be gases, liquids, or solids. They don't conduct heat or electricity. The solids are brittle, and they have no metallic luster. They readily accept electrons from metals to form salts. These include nitrogen, oxygen, fluorine, chlorine, bromine, and iodine.

Selenium is an element with a fascinating and multifaceted history that intricately links it to the evolution of chemistry and our understanding of nonmetallic elements. Its story unfolds from the early 19th century and flows into modern technological applications, showcasing its significance across various fields such as electronics, material science, and even biology. By diving deeper into the historical, chemical, and biological aspects of selenium, we can appreciate its remarkable journey from discovery to its contemporary uses in our daily lives.

Selenium was first discovered in 1817 by Jöns Jacob Berzelius, a prominent Swedish chemist who played a crucial role in developing modern chemistry. Berzelius isolated selenium while he was actually conducting research on tellurium. Remarkably, he observed that selenium shared several physical and chemical characteristics with tellurium, which prompted him to further investigate this intriguing element. The name "selenium" derives from the Latin "selenium," a nod to Selene, the Greek goddess of the moon, reflecting the element's spectral similarities to tellurium, named after Tellus, the Earth. This naming convention artfully symbolizes the connection between these two elements, both of which emanate a silvery sheen.

In the early decades following its discovery, selenium experienced a surge of interest from scientists keen on exploring its chemical reactivity and spectral properties. Berzelius and his contemporaries identified various oxidation states of selenium and examined its ability to form compounds with both metals and nonmetals. Among the striking features of selenium were its allotropes, particularly the red and gray forms, which quickly captivated the scientific community. The gray crystalline form is metallic-looking and exhibits fascinating semiconducting properties, while the red polymeric form is significantly less conductive, resembling a molecular configuration more akin to polymers than metals.

The late 19th century signaled a turning point for selenium, as its practical applications began to emerge. Scientists uncovered that selenium's ability to conduct electricity improves in the presence of light, leading to the development of light-sensitive devices that transformed how technology interacted with the natural world. One of the most notable innovations stemming from this discovery was the creation of selenium rectifiers and photoelectric cells, which utilized selenium's semiconductor-like behavior to convert light into electrical energy effectively. These early technologies would lay the groundwork for our modern understanding of photonics and semiconductor materials.

As the 20th century progressed, selenium's versatility continued to be explored, leading to its integration into a range of technological applications. In addition to its early roles in electricity and photography, selenium became increasingly relevant in the burgeoning fields of photocopying technologies, rectifiers, and solar cells. Its application in xerography, a critical process before the advent of modern laser technology, showcases how selenium played a significant role in advancing light-sensitive equipment. While silicon emerged as the dominant material in the field of semiconductors, selenium maintained its relevance by finding niche applications in various electronic devices and sensors.

Moreover, the existence of multiple allotropes of selenium highlights its complex nature. The gray form, recognized for its electrical conductivity and metallic appearance, contrasts with the red and black forms, which exhibit different properties due to their distinct molecular arrangements. This spectrum of allotropes allows for diverse applications across different industries, making selenium a unique element deserving of continued research.

On a biological frontier, selenium's importance cannot be overstated. Recognized as an essential trace element for many forms of life, including humans, selenium forms a foundational component of selenoproteins. These proteins are crucial for various biological processes such as supporting antioxidant defenses, facilitating thyroid hormone metabolism, and bolstering immune function. The recognition of selenium's significance in human health and nutrition surged in the 1950s and 1960s, marking a pivotal moment in our understanding of dietary element balances. Research has illuminated the fine line between deficiency and toxicity when it comes to selenium; inadequate intake can lead to conditions such as Keshan disease, while excessive selenium can result in selenosis, showcasing the element's dual nature.

In contemporary society, the quest for the right balance of selenium intake is focal to public health, influencing dietary recommendations and guidelines. Furthermore, environmental concerns regarding selenium highlights the risks posed by contamination due to agricultural runoff or mining operations. Such contamination can adversely impact water quality and ecosystems, necessitating rigorous monitoring and remediation efforts to protect biodiversity and human health.

The adaptability of selenium's semiconducting properties continues to incite interest in the fields of chemistry and material science, where researchers are actively investigating its potential applications in catalysis and energy storage. In photovoltaics and electronics, while silicon undeniably reigns supreme as the dominant material for solar cells, selenium-based compounds such as copper indium gallium selenide (CIGS) remain relevant in certain commercial technologies. This highlights selenium's position as a complementary element in the development of renewable energy technologies.

Finally, ongoing research into selenium's environmental behavior enhances our understanding of how it cycles through aquatic ecosystems and influences wildlife health. As scientists develop more precise analytical techniques to scrutinize selenium concentrations in various environments, our ability to monitor and remediate selenium presence evolves. This research not only aids in addressing current environmental challenges but also informs policy decisions aimed at protecting ecosystems affected by selenium contamination.

In conclusion, the narrative of selenium is one rich with discovery, innovation, and a deep connection to key scientific principles. From its initial isolation by Jöns Jacob Berzelius and its subsequent roles in early electrical technologies to its current status as an essential nutrient and material of interest, selenium's journey exemplifies the dynamic interplay of chemical elements and their impact on both industry and biology. As science continues evolving, it is likely that selenium will remain at the forefront of research, unlocking new potentials and solidifying its role in our lives for years to come.

Uses For Selinium

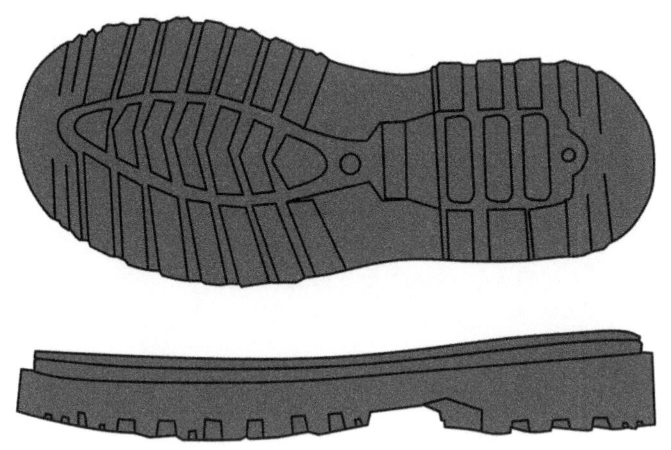

Selenium plays a crucial role in the rubber industry by being a key ingredient in the vulcanization process, which significantly enhances the properties of rubber. By incorporating selenium, manufacturers are able to create rubber that is not only stronger but also has a prolonged lifespan. This treatment allows the rubber to better maintain its shape, exhibit improved flexibility, and resist damage from various environmental factors, thereby ensuring that the final products meet quality standards and perform effectively over time.

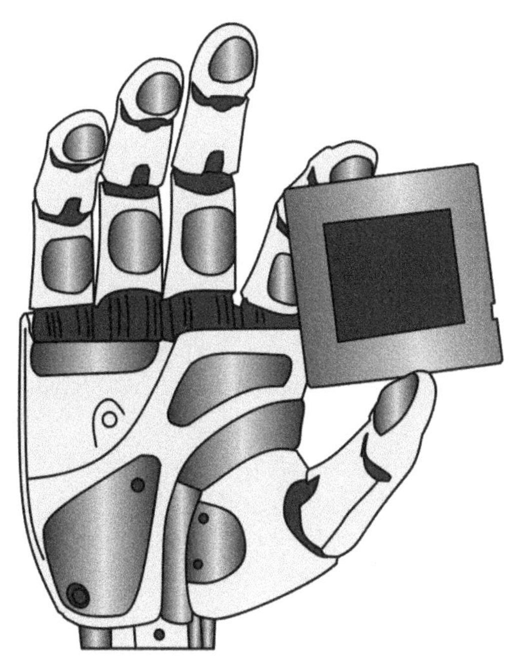

Selenium plays a crucial role in the development of hardware components for artificial intelligence systems. As a vital semiconductor material, selenium is utilized in various electronic devices, enabling efficient energy conversion and signal processing. Its unique properties, such as photoconductivity and temperature stability, make it particularly suitable for applications in sensors and solar cells, which are increasingly essential for powering AI technologies. By integrating selenium into the hardware components of AI systems, developers can enhance performance and reliability. This advancement supports the growing demand for more efficient data processing and machine learning capabilities, ultimately contributing to the evolution of smarter AI solutions.

Uses For Selenium

Selenium is an essential trace element that plays a critical role in plant growth and development. When soils are deficient in selenium, it can lead to reduced crop yields and compromised nutritional quality of the produce. By adding selenium to these selenium-deficient soils, farmers can significantly enhance crop nutrition. This micronutrient helps in various physiological processes, including photosynthesis and antioxidant defense, promoting overall plant health. Improved selenium levels can also lead to better yield and quality of crops, ensuring that they are not only more productive but also nutritionally beneficial for consumers, ultimately contributing to food security and public health.

Nano-Selenium-based materials have emerged as a promising solution for clean air filtering technologies. These innovative materials utilize the unique properties of selenium at the nanoscale to effectively capture and neutralize airborne pollutants. By integrating nano-selenium into filtration systems, researchers are able to enhance the efficiency and longevity of air purifiers. The high surface area and reactivity of nano-selenium allow for superior adsorption and catalytic properties, making it adept at removing hazardous substances such as volatile organic compounds (VOCs) and particulate matter. This advancement not only contributes to improved indoor air quality but also supports broader environmental efforts aimed at reducing air pollution.

Uses For Selenium

(Continued)

Selenium-containing biodegradable plastics represent a significant advancement in the quest for environmentally friendly materials. These innovative plastics not only decompose naturally, minimizing their impact on landfills and ecosystems, but also incorporate selenium, an essential trace element known for its antioxidant properties. This incorporation can enhance the durability and functionality of the plastics while simultaneously promoting environmental sustainability. As they break down, these materials contribute beneficial nutrients to the soil. Moreover, the use of biodegradable plastics reduces reliance on traditional petroleum-based plastics, offering a cleaner alternative that aligns with global efforts to combat pollution and promote a greener future for our planet.

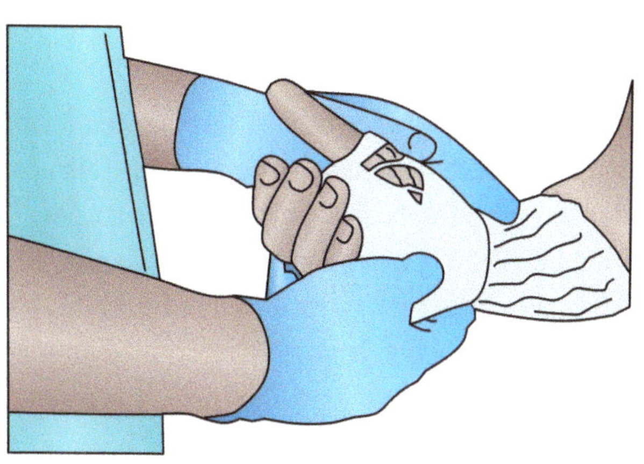

Tissue engineering, particularly through the incorporation of Selenium, plays a pivotal role in advancing regenerative medicine by supporting the growth of new tissues and organs. Selenium, a vital trace element, has been shown to enhance cellular functions and promote antioxidant activity, which is crucial for wound healing. By incorporating Selenium into various biomaterials used in tissue scaffolds, researchers can create an environment that stimulates cell proliferation and differentiation. This innovative approach not only accelerates the healing process but also improves the functionality of the regenerated tissues. Ultimately, the integration of Selenium in tissue engineering holds significant potential for improving patient outcomes in regenerative therapies.

The Source of Selenium

Selenium is a fascinating element that, unlike many others, is not found in nature in its pure, uncombined form. Instead, it occurs primarily as a compound, existing largely in association with various other elements. This characteristic places Selenium within a network of minerals, often embedded alongside metals such as copper, lead, and silver. One of the principal sources from which we extract Selenium is a mineral known as selenite. This mineral plays a pivotal role in the processes that allow us to acquire Selenium, a trace element used in various industries. Furthermore, seleniferous minerals—those specifically known to harbor Selenium—are essential sources that enrich our understanding of where this element can be found.

The journey of Selenium extraction begins with mining, which is the foundational step in accessing this unique element. When mining operations focus on extracting metals, especially copper, inadvertent extraction of Selenium occurs as well. The ore extracted from the Earth's crust often contains this valuable element. The refining of copper ore not only yields copper but also enables the separation and collection of Selenium. This process underscores the interconnected nature of the mining and metallurgy industries.

The Source of Helium (Continued)

Once the ores containing Selenium are obtained, they undergo smelting, a high-temperature process essential for extracting metals from their ores. Smelting involves heating the ore to high temperatures, which activates chemical reactions that yield the desired metals. Importantly, during this smelting process, the heat generates various byproducts, including Selenium. Indeed, it is from the residual materials, primarily in the form of selenium compounds known as selenides, that Selenium can be collected.

However, the journey of Selenium does not end with mere extraction. To isolate it in its elemental form, further processing is required. This involves the conversion of Selenium into its oxide state, which is then subjected to a series of intricate chemical reactions designed for purification. Through this careful methodology, the result is a shiny, reddish-black material that is visually distinct and possesses unique properties.

A particularly fascinating characteristic of Selenium is its ability to exist in multiple structural forms known as allotropes. Among these, the gray and red allotropes are the most notable. Gray selenium is recognized as the more stable form of the element and finds utility in an array of industrial applications, particularly in the realms of electronics and glass production. Its properties enable it to be used in solar cells, semiconductors, and photocopiers. Conversely, red selenium serves its unique purpose, often being incorporated into consumer products and further showcasing the versatility of this remarkable chemical element.

Beyond these industrial significances lies the integral role Selenium plays in biological systems across various forms of life, including humans. As a trace element, it operates as an antioxidant, assisting in protecting cellular structures from oxidative stress and damage inflicted by free radicals—harmful molecules that can lead to cellular dysfunction and disease. Selenium is also crucial for several vital biological functions, including metabolism and the synthesis of thyroid hormones. These hormones regulate numerous physiological processes in the body, demonstrating the element's importance in maintaining health.

Despite its vital roles, caution is warranted concerning Selenium intake. While essential in trace amounts, excessive exposure can lead to toxicity, resulting in adverse health consequences. Signs of selenium toxicity can range from gastrointestinal issues and hair loss to more severe complications such as neurological disorders. This fine balance underscores the importance of obtaining Selenium through a well-rounded diet consisting of selenium-rich foods like nuts, seeds, grains, and seafood. Contrary to common belief, not all Selenium intake is safe; therefore, understanding its role in our diet is crucial.

Geographically, Selenium can be found in various mining locations around the world. Notable regions include the United States, particularly the mining districts in places like Arizona, where copper mining activities often yield significant Selenium byproducts. Other significant sources are found in parts of China, Canada, and the Democratic Republic of Congo, as copper mining and mineral extraction in these regions also result in Selenium mining. The processing of Selenium typically occurs nearby the extraction sites, where ore is smelted, and further chemical processing takes place to yield the refined form of the element.

Exploring Selenium's multifaceted nature presents a comprehensive understanding of an element that affects both industrial applications and biological systems. From the earth where it is mined to the products in which it is utilized, Selenium underscores the intricate and interconnected relationships between nature, human health, and technological advances. Therefore, understanding Selenium's complex roles, alongside responsible consumption, highlights the delicate equilibrium between benefiting from nature's resources and maintaining our well-being.

Magical Elements of The Periodic Table

No Metal

No Magic

Actinium To Zirconium

Remember, "No Metal—
No Magic."
. . .And no technology.

Magical elementals from the Magical Elements of the Periodic Table books present all of the elements of the periodic table in fantastical and real life terms.

In the books, each elemental character has magical powers based on the properties of the elements that come from the land, air and water. They are the perfect group to introduce you to metals, metalloids, non-metals, halogens, noble gases and much more.

Unicorns, dragons, alchemists, knights, and goblins will show you how people of this world always have and always will depend upon the elements that our earth provides for all of our needs.

Use this Periodic Table as you would any other to spark an interest in the magical and real world properties of all the elements known today. You may be surprised at how prominently they feature in our every day lives.

1
H
hydrogen
Hildy
Textile Manufacturing

2
He
helium
Helha
Balloons

4
Be
beryllium
Berwyn
Musical Instrument

10
Ne
neon
Jalan
Advertising Signs

3
Li
lithium
Lillian
Batteries

9
F
Fluorine
Fleure
Strong Bones and Teeth

17
Cl
chlorine
Krysta
Swimming Pools

18
Ar
argon
Areg
Light Bulbs

12
Mg
Magnesium
Maggie
In Your Bones

8
O
oxygen
Ozzy
Air

16
S
sulfur
Xoe
Matches

34
Se
selenium
Selenita
Printers

36
Kr
krypton
Krypto
Detect Leaks

11
Na
sodium
Sorn
Salt

7
N
nitrogen
Nitra
Food Packaging

15
P
phosphorus
Phova
Fertilizer

33
As
arsenic
Arkyn
Poison

54
Xe
xenon
Xena
Used To Catch Species

5
B
boron
Boreas
Sports Equipment

6
C
carbon
Cole
Charcoal

14
Si
silicon
Silonar
Glass

32
Ge
germanium
Gemel
Camera Lense

52
Te
tellurium
Tellan
Vulcanized Rubber

13
Al
aluminium
Alumna
Airplanes

31
Ga
gallium
Gallant
LED Displays

50
Sn
Tin
Tinam
Liquid Crystal Display

51
Sb
antimony
Antz
Flame Resistant Fabric

30
Zn
Zinc
Dr. Zinko
Suntan Lotion

49
In
indium
Iker
Liquid Crystal Display (LCD)

53
I
Iodine
Jody
Cloud Seeding

29
Cu
copper
Cuprum
Money

48
Cd
cadmium
Cadmus
Power Tools

28
Ni
nickel
Nix
Guitar Strings

47
Ag
Silver
Silubhra
Ventilator

27
Co
cobalt
Cortss
Magnets

46
Pd
palladium
Paedin
Concert Flute

26
Fe
Iron
Town
Bicycle Chains

45
Rh
rhodium
Rovana
Finish for Jewelry

25
Mn
manganese
Mangar
Earth Movers

44
Ru
ruthenium
Ruth
Electrical Switches

24
Cr
chromium
Crowmist
Stainless Steel

43
Tc
technetium
Tephen
Radio Active Diagnosis

23
V
vanadium
Vana
Black Printer Ink

42
Mo
molybdenum
Maximo
Cutting Tools

22
Ti
titanium
Tilly
Aerospace

41
Nb
niobium
Nooah
Mag Lev Trains

21
Sc
scandium
Scandra
Bicycles

40
Zr
zirconium
Zora
Chemical Pipelines

20
Ca
calcium
Verity
Teeth

39
Y
yttrium
Yago
Microwave

19
K
Potassium
Pearl
Saline Drips

38
Sr
strontium
Strauna
Computer Screens

37
Rb
Rubidium
Ruby
Night Vision Glasses

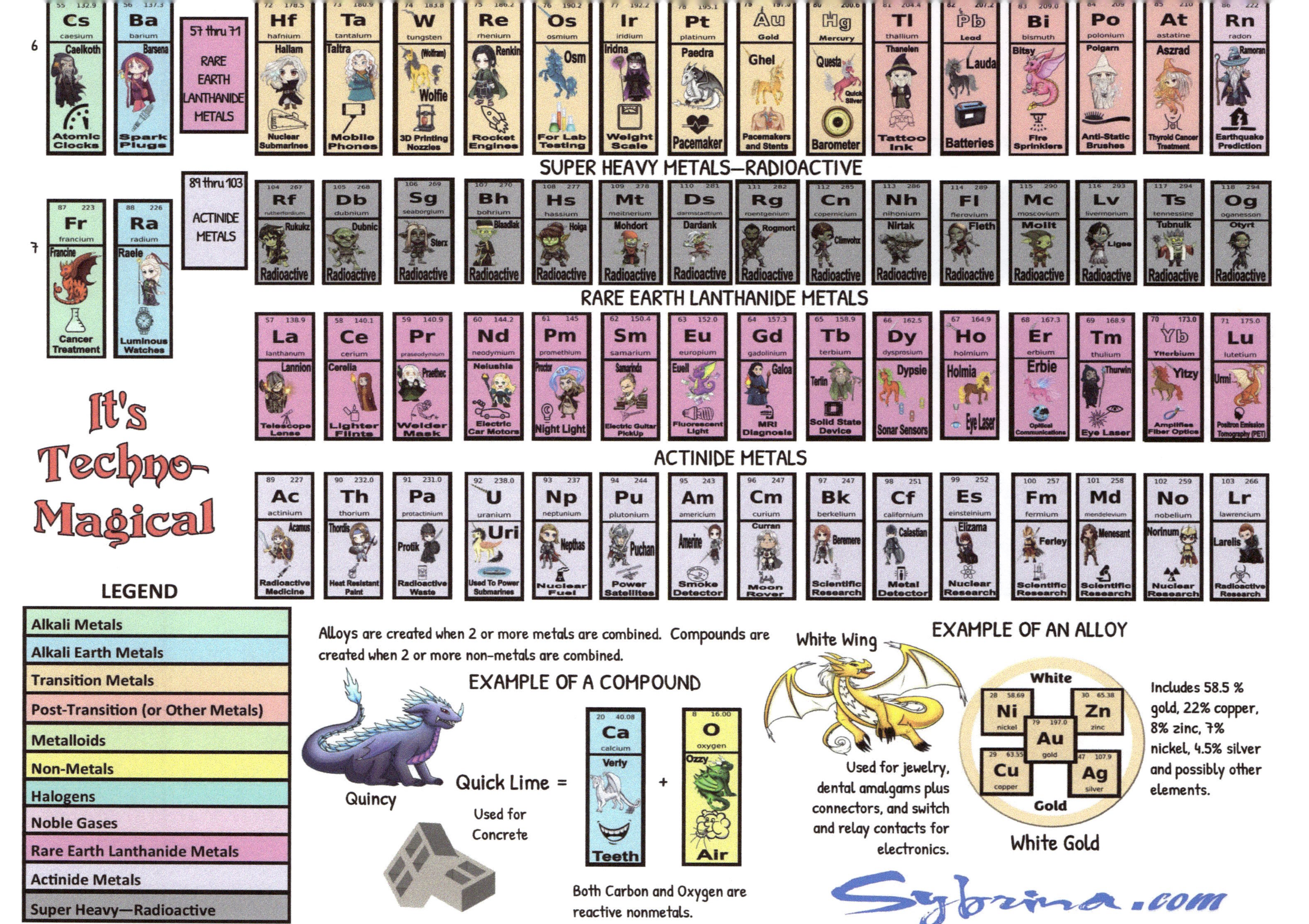
It's Techno-Magical

Cs 55 132.9 caesium Caelkoth Atomic Clocks
Ba 56 137.3 barium Barsena Spark Plugs
57 thru 71 RARE EARTH LANTHANIDE METALS
Hf 72 178.5 hafnium Hallam Nuclear Submarines
Ta 73 180.9 tantalum Taltra Mobile Phones
W 74 183.8 tungsten (Wolfram) Wolfie 3D Printing Nozzles
Re 75 186.2 rhenium Renkin Rocket Engines
Os 76 190.2 osmium Osm For Lab Testing
Ir 77 192.2 iridium Iridna Weight Scale
Pt 78 195.1 platinum Paedra Pacemaker
Au 79 197.0 gold Ghel Pacemakers and Stents
Hg 80 200.6 mercury Quick Silver Questa Barometer
Tl 81 204.4 thallium Thanelen Tattoo Ink
Pb 82 207.2 lead Lauda Batteries
Bi 83 209.0 bismuth Bitsy Fire Sprinklers
Po 84 209 polonium Polgarn Anti-Static Brushes
At 85 210 astatine Aszrad Thyroid Cancer Treatment
Rn 86 222 radon Ramoran Earthquake Prediction

Fr 87 223 francium Francine Cancer Treatment
Ra 88 226 radium Raele Luminous Watches
89 thru 103 ACTINIDE METALS

SUPER HEAVY METALS—RADIOACTIVE
Rf 104 267 rutherfordium Rukukz Radioactive
Db 105 268 dubnium Dubnic Radioactive
Sg 106 269 seaborgium Sterx Radioactive
Bh 107 270 bohrium Blaadlak Radioactive
Hs 108 277 hassium Hoiga Radioactive
Mt 109 278 meitnerium Mohdort Radioactive
Ds 110 281 darmstadtium Dardank Radioactive
Rg 111 282 roentgenium Rogmort Radioactive
Cn 112 285 copernicium Clinvohz Radioactive
Nh 113 286 nihonium Nirtak Radioactive
Fl 114 289 flerovium Fleth Radioactive
Mc 115 290 moscovium Molit Radioactive
Lv 116 293 livermorium Ligee Radioactive
Ts 117 294 tennessine Tubnulk Radioactive
Og 118 294 oganesson Otyrt Radioactive

RARE EARTH LANTHANIDE METALS
La 57 138.9 lanthanum Lannion Telescope Lense
Ce 58 140.1 cerium Cerelia Lighter Flints
Pr 59 140.9 praseodymium Praethec Welder Mask
Nd 60 144.2 neodymium Nelushla Electric Car Motors
Pm 61 145 promethium Proctor Night Light
Sm 62 150.4 samarium Samarinda Electric Guitar PickUp
Eu 63 152.0 europium Euell Fluorescent Light
Gd 64 157.3 gadolinium Galoa MRI Diagnosis
Tb 65 158.9 terbium Terlin Solid State Device
Dy 66 162.5 dysprosium Dypsie Sonar Sensors
Ho 67 164.9 holmium Holmia Eye Laser
Er 68 167.3 erbium Erbie Optical Communications
Tm 69 168.9 thulium Thurwin Eye Laser
Yb 70 173.0 Ytterbium Yitzy Amplifies Fiber Optics
Lu 71 175.0 lutetium Umi Positron Emission Tomography (PET)

ACTINIDE METALS
Ac 89 227 actinium Acamus Radioactive Medicine
Th 90 232.0 thorium Thordis Heat Resistant Paint
Pa 91 231.0 protactinium Protik Radioactive Waste
U 92 238.0 uranium Uri Used To Power Submarines
Np 93 237 neptunium Nephtas Nuclear Fuel
Pu 94 244 plutonium Puchan Power Satellites
Am 95 243 americium Amerine Smoke Detector
Cm 96 247 curium Curran Moon Rover
Bk 97 247 berkelium Beremere Scientific Research
Cf 98 251 californium Calastian Metal Detector
Es 99 252 einsteinium Elizama Nuclear Research
Fm 100 257 fermium Ferley Scientific Research
Md 101 258 mendelevium Menesant Scientific Research
No 102 259 nobelium Norinum Nuclear Research
Lr 103 266 lawrencium Larelis Radioactive Research

LEGEND
Alkali Metals
Alkali Earth Metals
Transition Metals
Post-Transition (or Other Metals)
Metalloids
Non-Metals
Halogens
Noble Gases
Rare Earth Lanthanide Metals
Actinide Metals
Super Heavy—Radioactive

Alloys are created when 2 or more metals are combined. Compounds are created when 2 or more non-metals are combined.

EXAMPLE OF A COMPOUND
Quincy
Quick Lime =
Ca 20 40.08 calcium Verly Teeth
+
O 8 16.00 oxygen Ozzy Air
Used for Concrete
Both Carbon and Oxygen are reactive nonmetals.

White Wing
EXAMPLE OF AN ALLOY
Used for jewelry, dental amalgams plus connectors, and switch and relay contacts for electronics.
White
Ni 28 58.69 nickel
Zn 30 65.38 zinc
Au 79 197.0 gold
Cu 29 63.55 copper
Ag 47 107.9 silver
Gold
White Gold
Includes 58.5% gold, 22% copper, 8% zinc, 7% nickel, 4.5% silver and possibly other elements.

Sybrina.com

Selenice The Wizard With The Selenium Staff

Symbol: Se Atomic Number: 34 Atomic Mass: 78.97

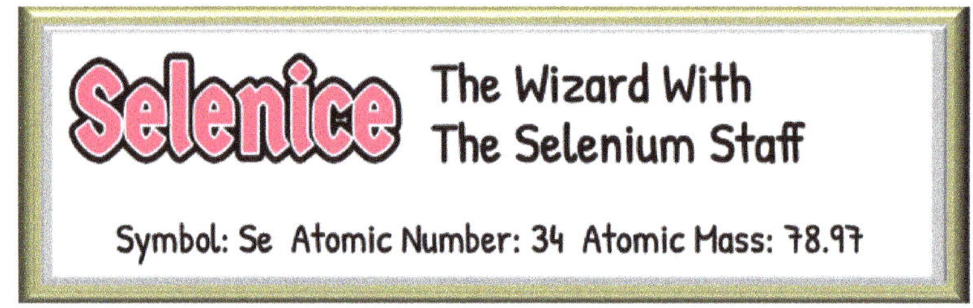

Selenium resides in Group 16 Period 4 on the Periodic Table.

The atomic symbol is Se. Its Atomic Number is 34. Its Atomic Mass is 78.97.

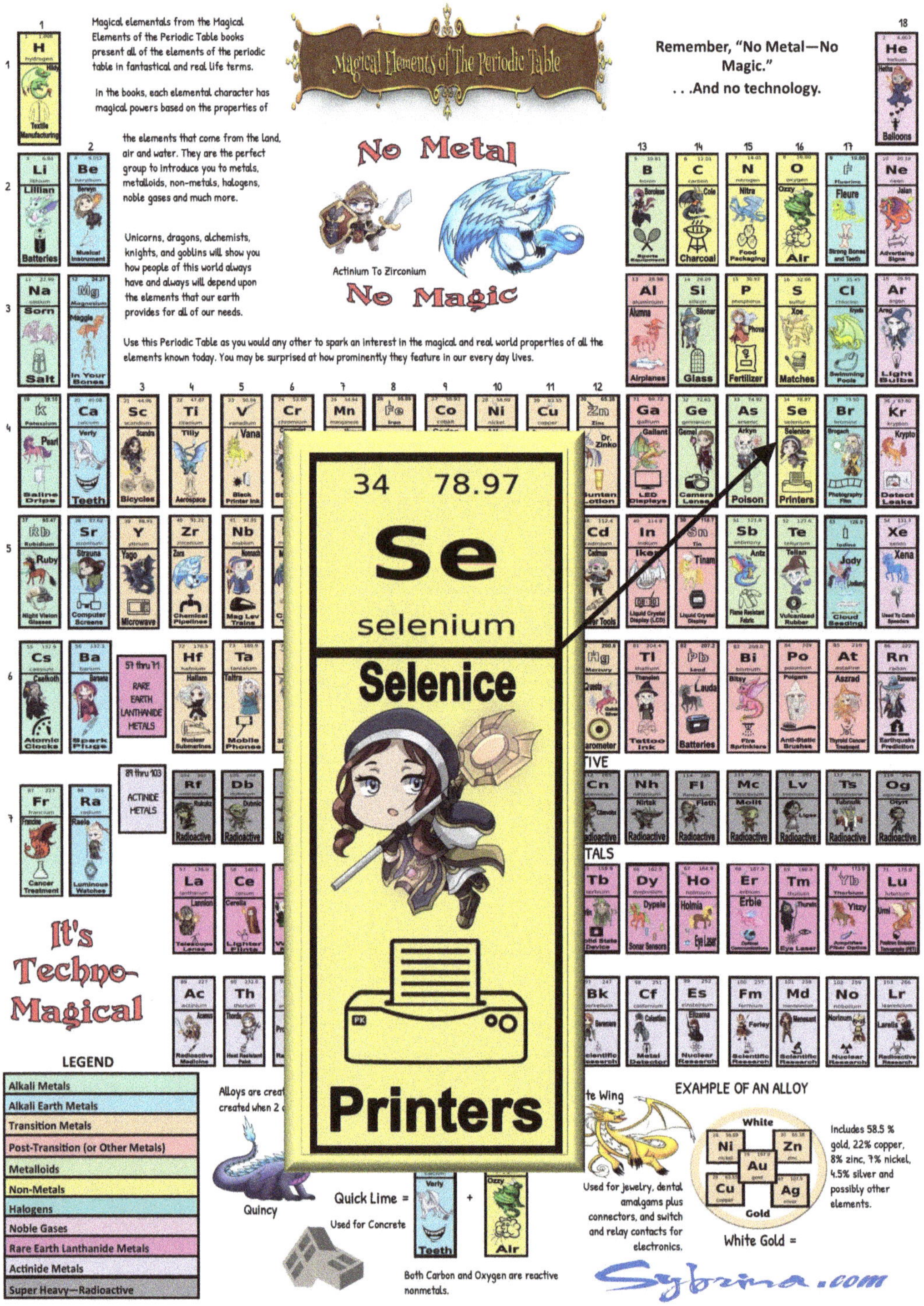

All Of The Periodic Table Elements Listed Alphabetically

ACTINIUM—*AC*—89

ALUMINUM—*AL*—13

AMERICIUM—*AM*—95

ANTIMONY—*SB*—51

ARGON—*AR*—18

ARSENIC—*AS*—33

ASTATINE—*AT*—85

BARIUM—*BA*—56

BERKELIUM—*BK*- 97

BERYLLIUM—*BE*—4

BISMUTH—*BI*—83

BOHRIUM—*BH*—107

BORON—*B*—5

BROMINE—*BR*—35

CADMIUM—*CD*—48

CALCIUM (Vital)—*CA*—20

CALIFORNIUM—*CF*—98

CARBON—*C*—6

CERIUM—*CE*—58

CESIUM—*CS*—55

CHLORINE (Keen)—*CL*—17

CHROMIUM—*CR*—24

COBALT—*CO*—27

COPERNICIUM—*CN*—112

COPPER—*CU*—29

CURIUM—*CM*—96

DARMSTADTIUM—*DS*—110

DUBNIUM—*DB*—105

DYSPROSIUM—*DY*—66

ERBIUM—*ER*—68

EINSTEINIUM-*ES*-99

EUROPIUM—*EU*—63

FERMIUM—*FM*—100

FLEROVIUM—*FL*—114

FLUORINE—*F*—9

FRANCIUM—*FR*—87

GADOLINIUM—*GD*—64

GALLIUM—*GA*—31

GERMANIUM—*GE*—32

GOLD—*AU*—79

HAFNIUM—*HF*—72

HASSIUM—*HS*—108

HELIUM—*HE*—2

HOLMIUM—*HO*—67

HYDROGEN—*H*—1

INDIUM—*IN*—49

IODINE *(JODIUM)* —*I*—53

IRIDIUM—*IR*—77

IRON—*FE*—26

KRYPTON—*KR*—36

LANTHANUM—*LA*—57

LAWRENCIUM—*LR*—103

LEAD—*PB*—82

LITHIUM—*LI*—3

LIVERMORIUM—*LV*—116

LUTETIUM (Unique)—*LU*—71

MAGNESIUM—*MG*—12

MANGANESE—*MN*—25

MEITNERIUM—*MT*—109

MENDELEVIUM—*MD*—101

MERCURY *(QUICK SILVER)* — *HG*—80

MOLYBDENUM—*MO*—42

MOSCOVIUM—*MC*—115

NEODYMIUM—*ND*—60

NEON (Jazzy)—*NE*—10

NEPTUNIUM—*NP*—93

NICKEL—*NI*—28

NIHONIUM—*NH*—113

NIOBIUM—*NB*—41

NITROGEN—*N*—7

NOBELIUM—*NO*—102

OGANESSON—*OG*—118

OSMIUM—*OS*—76

OXYGEN—*O*—8

PALLADIUM—*PD*—46

PHOSPHORUS—*P*—15

PLATINUM—*PT*—78

PLUTONIUM—*PU*—94

POLONIUM—*PO*—84

POTASSIUM—*K*—19

PRASEODYMIUM—*PR*—59

PROMETHIUM—*PM*—61

PROTACTINIUM—*PA*—91

RADIUM—*RA*—88

RADON—*RN*—86

RHENIUM—*RE*—75

RHODIUM—*RH*—45

ROENTGENIUM—*RG*—111

RUBIDIUM—*RB*—37

RUTHENIUM—*RU*—44

RUTHERFORDIUM—*RF*—104

SAMARIUM—*SM*—62

SCANDIUM—*SC*—21

SEABORGIUM—*SG*—106

SELENIUM— *SE*—34

SILICON—*SI*—14

SILVER—*AG*—47

SODIUM—*NA*—11

STRONTIUM—*SR*—38

SULFUR (Xanthous)—*S*—16

TANTALUM—*TA*—73

TECHNETIUM—*TC*—43

TELLURIUM—*TE*—52

TENNESSINE—*TS*—117

TERBIUM—*TB*—65

THALLIUM—*TI*—81

THORIUM—*TH*—90

THULIUM—*TM*—69

TIN—*SN*—50

TITANIUM—*TI*—22

TUNGSTEN—*W (WOLFRAM)* — 74

URANIUM—*U*—92

VANADIUM— *V*—23

XENON—*XE*—54

YTTERBIUM—*YB*—70

YTTRIUM—*Y*—39

ZINC—*ZN*—30

ZIRCONIUM—*ZR*—40

- Selenium derives its name from the ancient Greek word "selēnē," which translates to "moon." This nomenclature reflects its close relationship with tellurium, a chemical element named after the Latin term for earth, "tellus." The naming conventions for both elements illustrate their respective connections to celestial and terrestrial bodies.
- When pyrite extracted from the historic Falun Mine, renowned for being the first location where selenium was discovered, is subjected to burning, it produces a striking red solid. This remarkable transformation is accompanied by a pungent aroma reminiscent of horseradish, adding a unique olfactory characteristic to the chemical reaction.
- When it's dark, selenium exhibits poor conductivity for electricity, making it an inefficient conductor. However, when you shine a light on it, its ability to conduct electricity significantly improves, and this enhancement is directly proportional to the brightness of the light. Interestingly, selenium also possesses a unique property that allows it to convert light energy directly into electricity, showcasing its dual role in facilitating electrical conduction and energy transformation.

Did You Know? (continued)

- Alexander Graham Bell's first significant innovation following the invention of the telephone was the photophone. This groundbreaking device utilized the properties of selenium to transmit sound through light beams, creating a new method of communication. The photophone allowed voice waves to be converted into light waves, which could then be transmitted over distances. Bell's work on this technology marked a pivotal moment in the history of communication, highlighting his relentless pursuit of innovative solutions beyond the telephone.

- In 1904, Arthur Korn made a groundbreaking advancement by utilizing selenium cells to demonstrate the wireless transmission of photographs, an early precursor to what would eventually evolve into television technology. His innovative approach involved scanning images with selenium to convert them into electrical signals, which were then encoded onto a carrier wave for transmission. This pioneering work laid the foundation for future developments in wireless communication and imaging technology, showcasing the incredible potential of combining art with science.

- Just one Brazil nut exceeds the recommended daily intake (RDI) of selenium, which is set at 55 micrograms for adults. Eating an excessive number of Brazil nuts each day may lead to toxicity, with garlic breath being a well-known symptom resulting from certain volatile selenium metabolites. Selenium is a "double-edged sword" in human health that acts as a vital nutrient at low doses and a poison at high ones.

- The Daily Show with Jon Stewart featured a segment on June 14, 2012, that satirized the J.R. Simplot Company for polluting Southern Idaho waterways with selenium, leading to fish deformities like two-headed trout. The segment effectively brought national attention to a local environmental issue, using humor to highlight a serious case of corporate-driven pollution and a lack of regulatory control.

- In the 2001 sci-fi comedy, "Evolution", Selenium is a key plot element. In it, Nitrogen-based aliens are found to be vulnerable to selenium, leading to the use of dandruff shampoo (which contains selenium sulfide) to combat them.

- It has also been mentioned in several Star Trek Episodes, Isaac Asimov's short story "Runaround", a couple of episodes of Skinwalker Ranch, and in a Marvel Comics (Earth-616) story.

Structure Of Elements In The Periodic Table

Periodic tables are laid out in rows and columns.

Vertical columns are called Groups.

Each element is placed in a specific location because of its atomic structure. Elements are arranged in Families.

Horizontal rows are called Periods.

All rows read left to right.

The last two periods are part of periods 6 & 7.

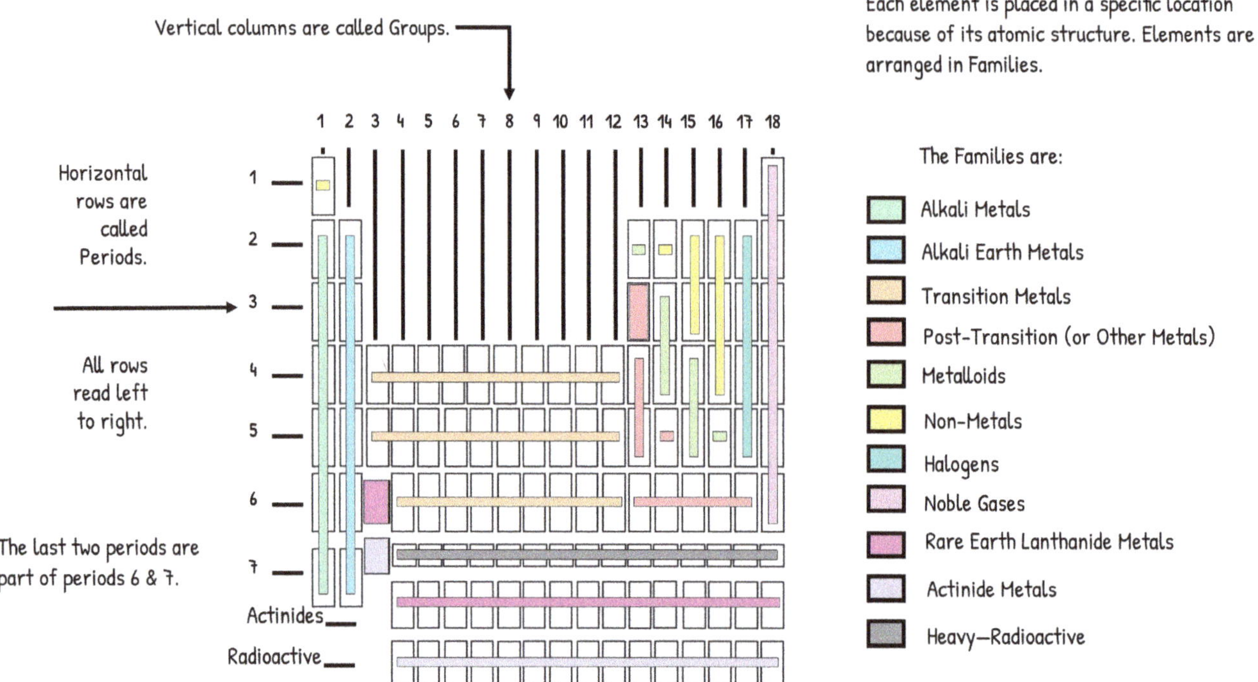

The Families are:

- Alkali Metals
- Alkali Earth Metals
- Transition Metals
- Post-Transition (or Other Metals)
- Metalloids
- Non-Metals
- Halogens
- Noble Gases
- Rare Earth Lanthanide Metals
- Actinide Metals
- Heavy—Radioactive

The term 'Element' is used to describe atoms with specific characteristics.
Every element in the first column or Group has 1 electron in the outer orbital (shell).
Every element in the second column (group two) has two electrons in the element's outer orbital.
The number designation of each Group represents the number of electrons in the element's outer orbital—
except for Group 18, Period 1—Helium. It only has 2 electrons.
Those electrons, called Valence Electrons, are what chemically bond with other elements.

Atomic Structure of Element: The atomic structure of an element refers to the arrangement of protons and neutrons in the nucleus of the atom, and the electrons in the electron cloud around the nucleus. Group 1, Period 1—Hydrogen is the only element that has no neutrons.

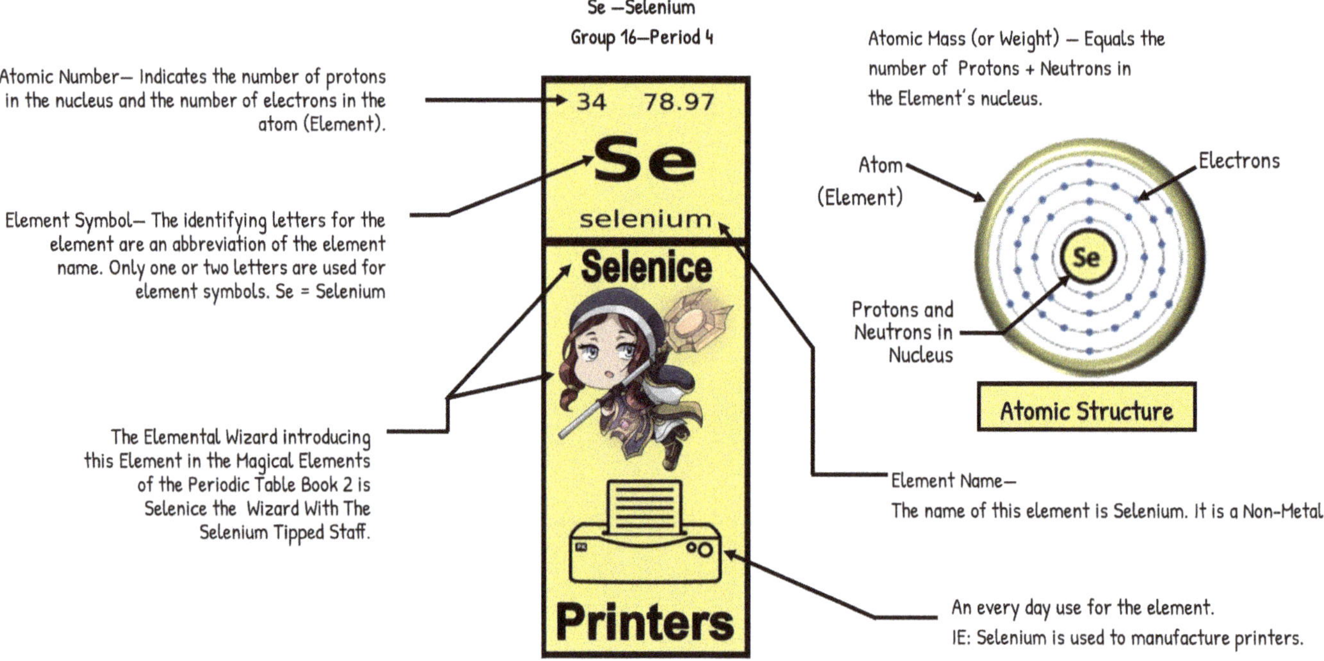

Atomic Number— Indicates the number of protons in the nucleus and the number of electrons in the atom (Element).

Element Symbol— The identifying letters for the element are an abbreviation of the element name. Only one or two letters are used for element symbols. Se = Selenium

The Elemental Wizard introducing this Element in the Magical Elements of the Periodic Table Book 2 is Selenice the Wizard With The Selenium Tipped Staff.

Atomic Mass (or Weight) — Equals the number of Protons + Neutrons in the Element's nucleus.

Element Name—
The name of this element is Selenium. It is a Non-Metal

An every day use for the element.
IE: Selenium is used to manufacture printers.

Types of Elements On The Periodic Table

Alkali Metals—Some metals on the periodic table are soft and shiny. They are so soft that they can be cut with a knife! These metals are excited to give away electrons to elements in need, making them highly reactive. Ther electron transfer creates a compound known as a salt. Surprisingly, these metals are not found in nature alone; they must be extracted from other sources. Examples of these metals include lithium, sodium, potassium, rubidium, cesium, and francium.

Alkali Earth Metals—The elements in column 2 of the periodic table have 2 outer electrons in their shell. Ther makes them very active with nonmetals that need electrons to stay stable. When they react, they make something called a salt. They are often found in nature all by themselves, and they can even conduct electricity. The elements are beryllium, magnesium, calcium, strontium, barium, and radium.

Post-Transition (or other Metals)— Elements directly to the right of the transition metals. They are known as "poor metals: and are soft and brittle. These include aluminum, gallium, indium, tin, thallium, lead, bismuth, zinc, cadmium and mercury.

Transition Metal—The main metals are found in the middle and bottom rows of the periodic table. They look like metal, can conduct electricity, can bend and be shaped easily. The period 4 transition metals are scandium, titanium, vanadium, chromium, manganese, iron, cobalt, nickel, copper, and zinc. The period 5 transition metals are yttrium, zirconium, niobium, molybdenum, technetium, ruthenium, rhodium, palladium, silver, and cadmium. The period 6 transition metals are lanthanum, hafnium, tantalum, tungsten, rhenium, osmium, iridium, platinum, gold, and mercury. The period 7 transition metals are the naturally-occurring actinium, and the artificially produced elements rutherfordium, dubnium, seaborgium, bohrium, hassium, meitnerium, darmstadtium, and roentgenium.

Metalloids—The elements called metalloids are a mix of metals and nonmetals. They look like metals, but can't conduct electricity very well. They also break easily and act like nonmetals. These include boron, silicon, germanium, arsenic, antimony, tellurium, astatine, and polonium.

Non-Metals—These elements reside in columns 15-17, and can be gases, liquids, or solids. They don't conduct heat or electricity. The solids are brittle, and they have no metallic luster. They readily accept electrons from metals to form salts. These include nitrogen, oxygen, fluorine, chlorine, bromine, and iodine.

Halogens—Halogen chemicals are a special type of element. When they mix with metal, they become a kind of salt. Halogens are super reactive because they like to take an electron from metals. They can be found in column 17 of the element table. Some of them can be found in nature, but most are very dangerous and can hurt you if you touch them. They include fluorine, chlorine, bromine, iodine, and the radioactive elements astatine and tennessine.

Noble Gases—These elements reside in column 8. They are all odorless, colorless gases that are chemically very stable (inert). They don't generally form compounds by bonding with another element. These include helium, neon, argon, krypton, xenon, and radon.

Lanthanide Rare Earth Minerals—The Japanese call them "the seeds of technology." The US Department of Energy calls them "technology metals." These elements have atomic numbers 57-71. They are vital to industry. They can be added to metals to strengthen them to make alloys such as stainless steel, used to refine crude oil, and are crucial in producing technology—electronics, telecommunications, and metal devices to name a few. They are lanthanum, cerium, praseodymium, neodymium, promethium, samarium, europium, gadolinium, terbium, dysprosium, holmium, erbium, thulium,

Actinide Metals—Any of a series of chemically similar metallic elements with atomic numbers ranging from 89 (actinium) to 103 (lawrencium). All of these elements are radioactive, and two of the elements, uranium and plutonium, are used to generate nuclear energy. The lanthanides and actinides are sometimes called the inner transition metals, referring to their properties and position on the table. They are actinium, thorium, protactinium, uranium, neptunium, plutonium,

Super Heavy—Radioactive—Superheavy elements are those elements with a large number of protons in their nucleus. Elements with more than 92 protons are unstable; they decay to lighter nuclei with a characteristic half-life. They do not occur in large quantities (if at all) naturally on earth, and only exist briefly under highly controlled circumstances. They include lawrencium, rutherfordium, dubnium, seaborgium, bohrium, hassium, meitnerium, darmstadtium, roentgenium, copernicium, nihonium, flerovium, moscovium, livermorium, tennessine, and oganesson.

Alloys

An **alloy** is a mixture of chemical elements of which at least one is a metal. An alloy is a solid. Unlike chemical compounds with metallic bases, an alloy will retain all the properties of a metal in the resulting material, such as electrical conductivity, ductility, opacity, and luster, but may have properties that differ from those of the pure metals, such as increased strength or hardness. In some cases, an alloy may reduce the overall cost of the material while preserving important properties. In other cases, the mixture imparts synergistic properties to the constituent metal elements such as corrosion resistance or mechanical strength. Some of the most common alloys are

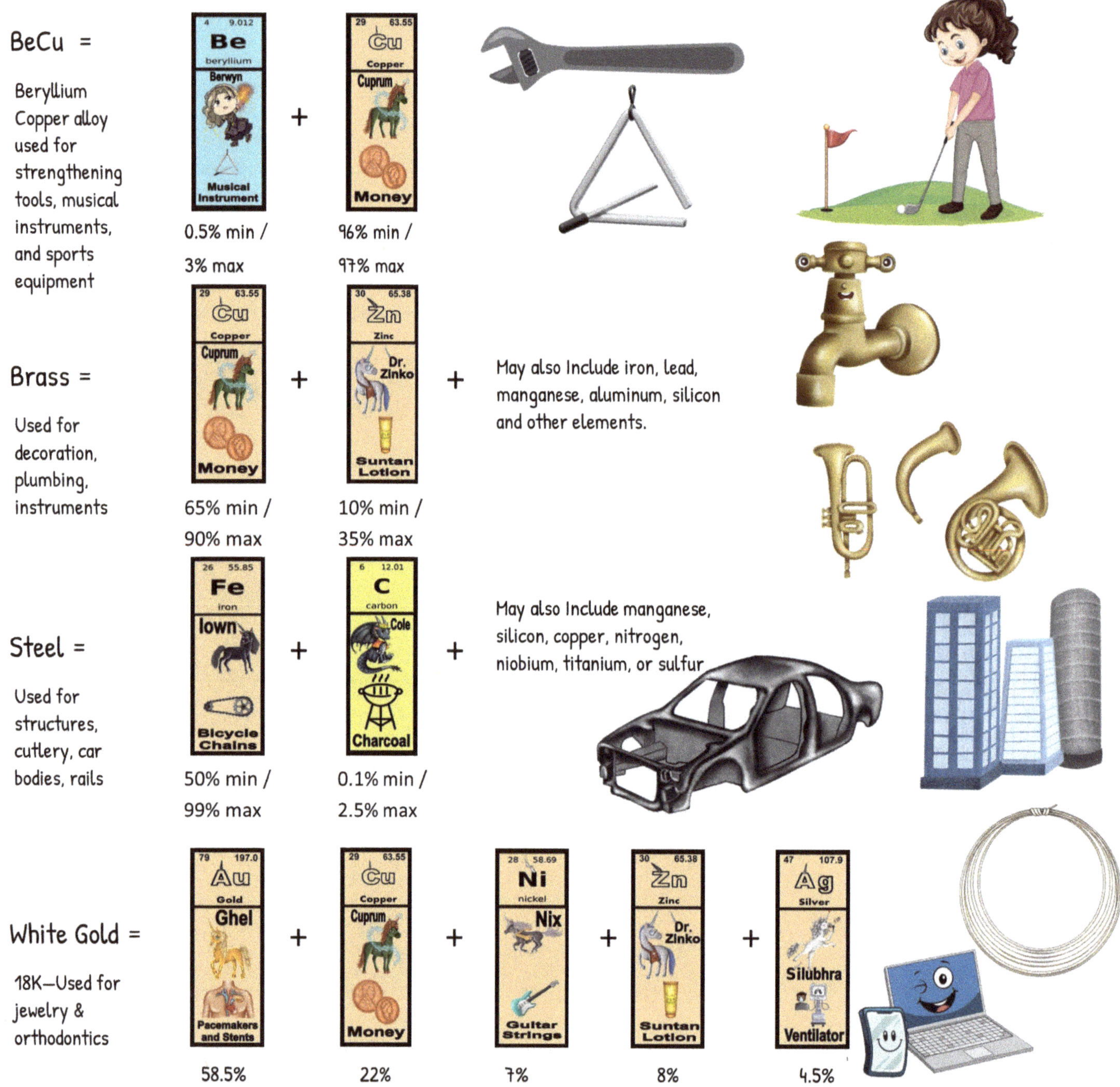

Some other common alloys are Bronze, Cast Iron, Cupronickel, Magnalium, Mischmetal, Nichrome, Nitinol, Pewter, Solder, Sterling Silver and Tungsten Carbide.

The above chart only shows a few of the hundreds of metal combinations. For instance, 24 carat gold is a pure naturally occurring yellow metal. There are four basic shades of gold alloys: yellow gold, white gold, rose gold, and green gold. A huge range of other colored golds are also possible, including red (gold and copper), grey (gold, iron and copper), purple (gold and aluminum), blue (gold and iron) and black (gold and cobalt), depending on the amounts of different metals alloyed together.

Compounds

A **compound** is a substance formed due to the chemical union (a chemical reaction) between two or more atoms or molecules. Ideally there should be two or more elements to form a compound. Most of the compounds are from a non-metallic origin.

Fluorocarbon =

Used for— waterproofing agents, lubricants, sealants, and leather conditioners.

 + 

Both Carbon and Fluorine are reactive nonmetals.

Sodium Fluoride =

Used for—the fluoridation of drinking water, in toothpaste, in metallurgy, and as a flux, and is also used in pesticides and rat poison.

 + 

Sodium Fluoride is a simple ionic compound, made of the sodium (Na+) cation and fluoride (F-), an anion of Flourine.

Potassium Iodide =

It's medical use is to block absorption of radioactive iodine by the thyroid gland.

 + 

Potassium iodide (also called KI) is a salt of stable (not radioactive) iodine. Potassium is an alkali metal and Iodine is a reactive nonmetal.

Calcium Bromate =

It is used as a bread dough and flour "improver" or conditioner

 + 

Calcium is an alkaline earth metal and Bromine is a reactive nonmetal.

Can you guess the most important compound of all?

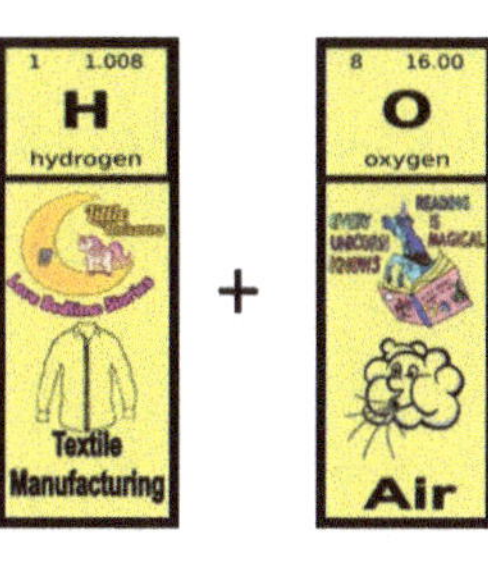

= ??????

(Answer can be found below.)

The above chart only shows a few of the millions of combinations of elements that create compounds.

Iodine, alone has 37 compounds. Some of them are Ammonium iodide (NH4I), Cesium iodide (CsI), Copper(I) iodide (CuI), Hydroiodic acid (HI), Iodic acid (HIO3), Iodine cyanide (ICN), Iodine heptafluoride (IF7), Iodine pentafluoride (IF5), Lead(II) iodide (PbI2), Lithium iodide (LiI), Nitrogen triiodide (NI3), Potassium iodate (KIO3), Potassium iodide (KI), Sodium iodate (NaIO3), and Sodium iodide (NaI).

Hydrogen is used in more compounds (nearly 100) than any other element because it can form bonds with almost all metals, metalloids, and non-metals. Some of the most common are water (H2O), Hydrogen Peroxide (H2O2) and table sugar (C12H22O11).

Neon forms no known compounds.

Definitions

Atomic Structure of Element: The atomic structure of an element refers to the arrangement of protons and neutrons in the nucleus of the atom, and the electrons in the electron cloud around the nucleus.

Atomic Number: An element's atomic number refers to the number of protons it has in its nucleus. In a neutral atom the number of protons always equals the number of electrons.

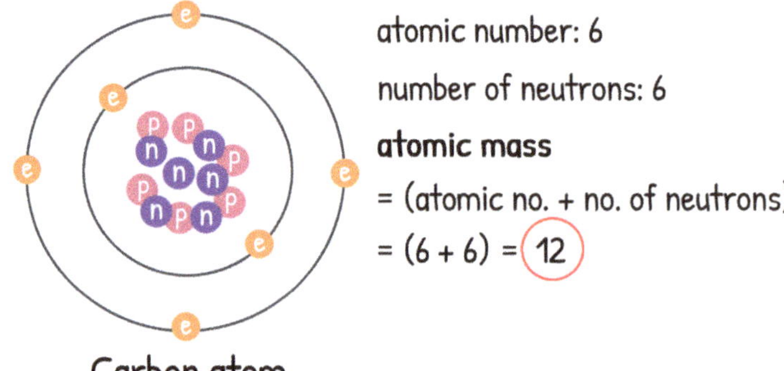

Atomic Weight (Mass) of Element: The atomic mass of an element is how heavy it is. It is made up of protons and neutrons that are in the middle of the element. Some elements have different versions with different amounts of neutrons, but they still have the same amount of protons. The atomic mass is the average weight of all these versions of the element.

Allotrope: Allotropes are different forms of an element that look and act different, but are made of the same stuff. Some elements have more than one form. For instance, carbon can be a shiny diamond or a gray pencil lead called graphite.

Isotope: Isotopes are different types of atoms that have the same parts, like protons and electrons, but they have a different number of neutrons. For example, the three most stable isotopes of hydrogen: protium (A = 1), deuterium (A = 2), and tritium (A = 3).

Crystalline Structure of Element: The crystalline structure of an element is how its atoms, ions, or molecules stick together in a pattern to make a cool crystal shape.

Ferrous and Non-Ferrous Metals: When we say ferrous metal, it means that iron is a big part of the metal. But if there's only a little bit of iron in the metal, we call it non-ferrous. The word "ferrous" comes from Latin and means iron, which is why iron's symbol is Fe.

Ductile Metals: These are capable of being made into long, thin wire or thread. Copper and Silver are ductile metals.

Malleable Metals: These can be hammered or rolled into thin sheets without cracking or breaking. Gold is malleable.

Ferromagnetic: Materials that are strongly attracted to a magnet. Such materials can be permanently magnetized. These include the elements iron, nickel and cobalt and their alloys, some alloys of rare-earth metals, and some naturally occurring minerals such as lodestone.

Magnetostriction—Ther is the term for a special thing that happens to magnetic materials. When these materials get turned into magnets, they also change their shape or size.

Paramagnetic: Slightly attracted to a magnetic field, but do not retain magnetic properties once the field is removed.

Diamagnetic: Slightly repelled by a magnetic field, but do not retain magnetic properties once the field is removed.

Electrical Properties: Conductor—a thing that lets electricity flow through it. Semi-conductor—a special material (usually silicon) that can conduct electricity, but not as well as metal. Insulator (non-conductor)—a material (usually glass) that stops electricity from flowing.

Reactive Gas: These gases are really good at reacting with stuff! They are called "sticky gases" because they can react to things like plastic and wet surfaces when they touch them. These are nitrogen, oxygen, hydrogen, carbon dioxide, fluorine, and chlorine.

Non-Reactive Gas: An inert gas is like a super shy gas that doesn't like to hang out with other chemicals. It doesn't make any new friends by reacting with them, so it doesn't form any chemical compounds. We also call these special gases "noble gases."

What Makes Selenium Seem Magical?

Selenium, a chemical element designated by the symbol "Se" on the periodic table, often evokes intrigue and curiosity. To the modern eye, it is simply one of the many elements that comprise the earth and play distinct roles in both nature and industry. However, when we view Selenium through the lens of magic, it transforms from a mere component of the earth into something that feels almost enchanting and mystical. This fascinating element possesses properties that can be perceived as magical, connecting our world to the wonders of chemistry and the marvels of nature.

To understand Selenium's magic, we must first delve into its versatility in nature. Selenium exists in various forms, showcasing a non-metallic, crystalline structure as well as an amorphous powder. This ability to change its appearance suggests an intriguing fluidity, reminiscent of the shape-shifting abilities often attributed to mystical creatures in folklore. Much like a wizard might transform into a different being, Selenium adapts and changes, embodying the beauty of transformation that mirrors certain aspects of our own lives. To the ancients and those who lived before the comfort of modern science, such transformative qualities would have sparked awe and wonder, potentially leading to myths around this element as a sort of divine gift or magical substance, given the lack of understanding of molecular structures and elemental properties at the time.

One of the most captivating aspects of Selenium is its role in biological systems. It has been established as an essential trace element required by many living organisms, including humans. In small amounts, Selenium can have potent effects on health, similar to how a spell might enhance one's abilities. This almost magical essence aids various bodily functions, particularly in its role as a powerful antioxidant that helps protect cells from damage. This supportive role in our health could easily be interpreted as a form of magic; after all, what could be more mystical than the sustenance of life and vitality relying on a humble element found in the earth?

Interestingly enough, during the pre-20th century era, the notion of health and illness was often intertwined with supernatural beliefs and phenomena. People relied heavily on the interpretation of their environments and the elements within them. If Selenium had been known during these times, it might have been viewed with reverence, likely attributed to divine forces or the mystical properties of nature. Herbs, minerals, and compounds were often deemed to have healing powers, and selenium could have found its place in these ancient remedies and potions as a magic mineral believed to hold the key to vibrant health and longevity.

What Makes Selenium Seem Magical? (Continued)

In addition to its biological importance, Selenium is known for its unique interactions with other elements. It forms compounds that are essential for various industrial applications, including electronics and glassmaking. When used in electronics, Selenium plays a pivotal role in photocells and photovoltaic devices, harnessing light and converting it into energy. This transmutation of light into power may have evoked feelings of enchantment in earlier societies who observed how simply harnessed elements could yield such powerful energy. In fairy tales, one might imagine a spell being cast to channel energy, reflecting how Selenium channels power in our physical world, bridging the divide between the realms of magic and science.

Selenium's electrical conductivity is another captivating facet of its character. It demonstrates duality, behaving like a metal at times—efficiently conducting electricity—while at other moments is an insulator. Such shifting qualities could easily remind one of a sorcerer's ability to conjure spells at will—sometimes mysterious and hidden, while at other times powerful and apparent. As the ancients sought to understand the stars and the forces of nature, they might have perceived Selenium's properties as a considerable reflection of the human experience, where change and flux are constant.

Moreover, the discovery of Selenium itself carries an aura of magic. This element was first identified in 1817 by the Swedish chemist Jöns Jacob Berzelius, who marveled at its remarkable properties. Known for his curiosity and insight, Berzelius can be likened to a modern-day alchemist unveiling secrets long concealed. Much like wizards who delve into ancient tomes and arcane rituals, chemists engage deeply with elements like Selenium in an effort to uncover their potential. Had Berzelius lived a century earlier, he may have been regarded as a mystical figure, someone who communicated with the elements and brought forth new understandings of the world through a form of scientific sorcery.

Finally, Selenium's role in the environment should not be overlooked. Naturally present in soil, water, and certain foods, its existence serves as a constant reminder of its connection to our surroundings. In folklore, many magical beings are tied to the land and nature, just as Selenium is part of the interconnectedness that binds all life. Whether nestled in a fertile patch of soil or found within the very foods we consume, Selenium embodies the idea of guardian magic, sustaining life and reinforcing the bond between humanity and nature.

In conclusion, viewed through a magical lens, Selenium reveals a world of wonder. Its ability to transform, its vital role in health, and its captivating properties underscore the beauty of science within nature. Selenium, in all its forms and functions, illustrates how even the most ordinary elements can possess extraordinary qualities. As we explore this enchanting element, we find that the magic we seek is not always found in mythical creatures or ancient spells, but rather woven into the fabric of the very elements that surround us. In this way, Selenium serves as a reminder that magic exists in the scientific marvels of our world, waiting patiently to be understood and appreciated.

Meet Selenice, The Wizard With The Selenium Tipped

The valley wore two faces that day: one old as the hills, one bright with screens and buzzing devices that hummed like distant bees. In the town of Brindlewick, where thatched roofs met glass towers, the clash of eras was not a battle but a conversation, each side listening for the other's tune. At the heart of Brindlewick's square stood a figure both familiar and uncanny: Selenice. This wizard exudes a charming blend of courage and cautious curiosity. Small in stature but mighty in presence, she drapes her outfit in regal purples, golds, and moonlit whites, hinting at a lineage of arcane tradition. Her hooded cloak frames a determined face, with wide, expressive blue eyes that gleam with wonder and a hint of mischief. A neatly braided chestnut braid peeks from beneath the hood, suggesting practicality and discipline.

Her staff is a gleaming instrument of power: a long silver shaft topped by an ornate, sun-kissed headpiece that radiates authority and focus. The staff's design—the geometric gold motifs and the central gem—speaks of precise, methodical magic rather than flashy showmanship. She grips it with confident, gloved hands that signal readiness for spells, wards, or whimsical experiments. She moves with a light, almost playful agility, as pink petals or sparkles swirl around her, enhancing the sense that magic is both delightful and delicate in her care. Selenice is a steadfast guardian of wonder.

But her power extends beyond the old textbooks and mythic sigils. The town's new age infrastructure—fiber lines, municipal drones, solar ferries gliding along the river—bounds Brindlewick into a web of circuits and signals. Selenice had learned to walk both pathways at once: the ancient path of incantation and the modern route of code and circuitry. She easily controls computers, smartphones, and other cool gadgets. With her powers, she can do things like change information, mess up computer systems, or protect against tech dangers. She can jump into digital worlds, move data around, and block bad electronic stuff. Yet she never wields power recklessly; she tempers the lust for control with the discipline of an old-world mage who remembers every spell invites a price.

The first test came on a fog-veiled morning when Brindlewick's river, a lifeline for mills and markets, stuttered to a halt. The water's gleam dulled to pewter; fish paused, and a chorus of worry rose from the docks where merchants counted their wares as if counting their breaths. The mayor—a pragmatic woman with copper wire spectacles and a smile that had weathered many storms—called for help. Windows in the municipal building flickered with alert icons: a cluster of emojis that looked like tiny thunderstorms. The town's old clock tower rang once, twice, then paused, as if listening.

Selenice stood at the square's edge, her staff catching the last light of dawn, a sun-kissed crown of gold at its head. The pink petals that swirled around her paused mid-air, as if listening, too. She spoke with that calm cadence that made even the most frantic heart slow. This is not a show, she reminded them in a voice that could travel across centuries and also across networks. This is a conversation, a negotiation between what has always been and what must be.

Her staff hummed with a quiet resonance, not the roar of a dragon, but the steady heartbeat of a doe—listen first, act second. The central gem pulsed in time with the river's own heartbeat, which felt almost like an echo from a distant memory. She asked a simple question, one that bridged old wisdom and modern need: Where does this murk begin, and where does it end? Then she did what a good guardian does—listen first, act second.

The town's engineers uploaded a map of the river's undercurrents to a public display. Selenice touched the data with her staff, and the geometric gold motifs sparked into light, tracing the water's hidden routes as if laying a map on the air itself. She blended spellcraft with a digital spellbook, a hybrid charm that could coax the river to reveal its burdens without harm. The central gem glittered with a cool, patient light as she summoned wards not only on the water itself but on the flow of information around it—protective filters for the sensors and safeguards for the grid that connected the mills to the market.

"This is not merely a restoration," she told the crowd, and her blue eyes flashed with that gleam of wonder and mischief she kept so close to her heart. "It is a reweaving—an interlacing of streams both literal and virtual. The river's spirit is a current of memory; the city's memory is stored in code and stone. We must honor both." The gold motifs on her staff traced sigils into the air, and the central gem pulsed with a rhythm that felt like a chorus—old chants meeting new melodies.

The river's murk began to loosen, but Selenice did not stop there. She extended her influence into Brindlewick's digital heart—the municipal servers, the clinics, the street lights that now hummed with an ever-so-slight blue glow. The staff's power shifted from healing water to safeguarding information. She could, if she chose, push data into or out of the network, alter the course of digital weather, or set up barriers against malevolent intrusions. But she used that potential as a precaution, never a temptation for domination. She explained to the gathered townsfolk that true harmony required balance: the old world's reverence for temperance and the new world's appetite for speed and connectivity.

In an era when cyber-spectral threats hovered like digital phantoms, Selenice offered an alliance. She trained a cohort of young coders who wore cloaks—of a more modern fabric, perhaps, but still respectful of ritual and discipline. They learned to code with the same patience she showed when mapping sigils in the air or tracing circles on the cobbles. She told them that information was a living thing that needed care, just as water or soil did. The central gem reminded them of the responsibility that came with knowledge: the power to heal, to protect, and to shape futures, but never to crush.

The town's market gained a new heartbeat as well. Vendors reported fewer disruptions from weather shifts or power glitches; the loom-house could run on solar energy longer into the day; the herbalist's apothecary obtained precise data about pollen counts and soil pH. Brindlewick began to orbit around a single ideal: that wonders could be woven from the threads of old world prudence and modern world ambition. The pink petals continued to swirl around Selenice, ethereal and gentle, a reminder that magic could be both delightful and delicate in her care—and that even the most advanced gadgetry could benefit from the human touch of patience and care.

Some nights, when the river's silver skin reflected the moon, Selenice stood at the square's edge, the staff resting easily in her gloved grip. The townsfolk spoke of her like a guardian who could speak with both ancient spirits and silicon sprites. A child asked her how she kept so many worlds from colliding. She knelt to the child's level, the neatly braided chestnut braid visible beneath the hood, practical and calm. "The worlds won't collide if we remember to listen," she said, and the girl's eyes widened with the same wonder and mischief that Selenice carried. The lesson was simple yet profound: courage must be tempered by curiosity, and curiosity by restraint.

As harvest season rolled in, Brindlewick celebrated not only abundance but a new kind of wealth—the shared knowledge of a community that had learned to balance its roots with its wings. A small library of the town's history—old tomes in leather, new tablets in glass—sat side by side beneath a canopy of stars. Selenice's staff stood propped against a bench where the mayor and the town's elders discussed architecture and ethics, their conversation drifting into visions of a school that taught both rune-scribing and programming. The school would be a place where apprentices learned to fix a river's memory as deftly as they learned to fix a broken code, where wards and firewalls worked in tandem, not in tension.

"This is more than magic," Selenice said in a late-night gathering, the pink petals still orbiting the lamp-lit square. "It is stewardship across centuries, a fusion of the old world's reverence and the new world's invention. The staff's selenium-based power is a metaphor and a tool—a reminder that the world's energy, like the element selenium, can be a spark of health and protection if it is kept under careful stewardship." She paused, letting the syllables settle like dust motes in a sunbeam. "We will guard our people, our rivers, and our networks with kindness, precision, and courage."

The townsfolk nodded, their faces lit by the glow of lanterns and lit screens alike. The fusion of old and new had become their living creed. Selenice, the small wizard with a mighty presence, moved through the square with that familiar grace—the hooded cloak, the blue eyes that gleamed with wonder, the chestnut braid peeking from beneath, the staff with its sun-kissed headpiece and central gem. She did not merely perform wonders; she built a bridge between times, teaching a community to trust its own power and to use it with care.

And so Brindlewick thrived, a beacon where the ancient arts and modern marvels walked hand in hand. The town learned that magic could be practical, that technology could be enchanted, and that a single, steadfast guardian of wonder could guide both systems toward harmony. Selenice's blend of courage and caution remained their compass, a reminder that true wonder lies not in dominance over the world but in the art of listening, learning, and lifting one another into a brighter dawn. The petals drifted, the screens glowed, and the valley held its breath, knowing that the next wonder was only a quiet, deliberate breath away.

Enjoy This Coloring Page Featuring

Selenice The Wizard With The Selenium Tipped Staff

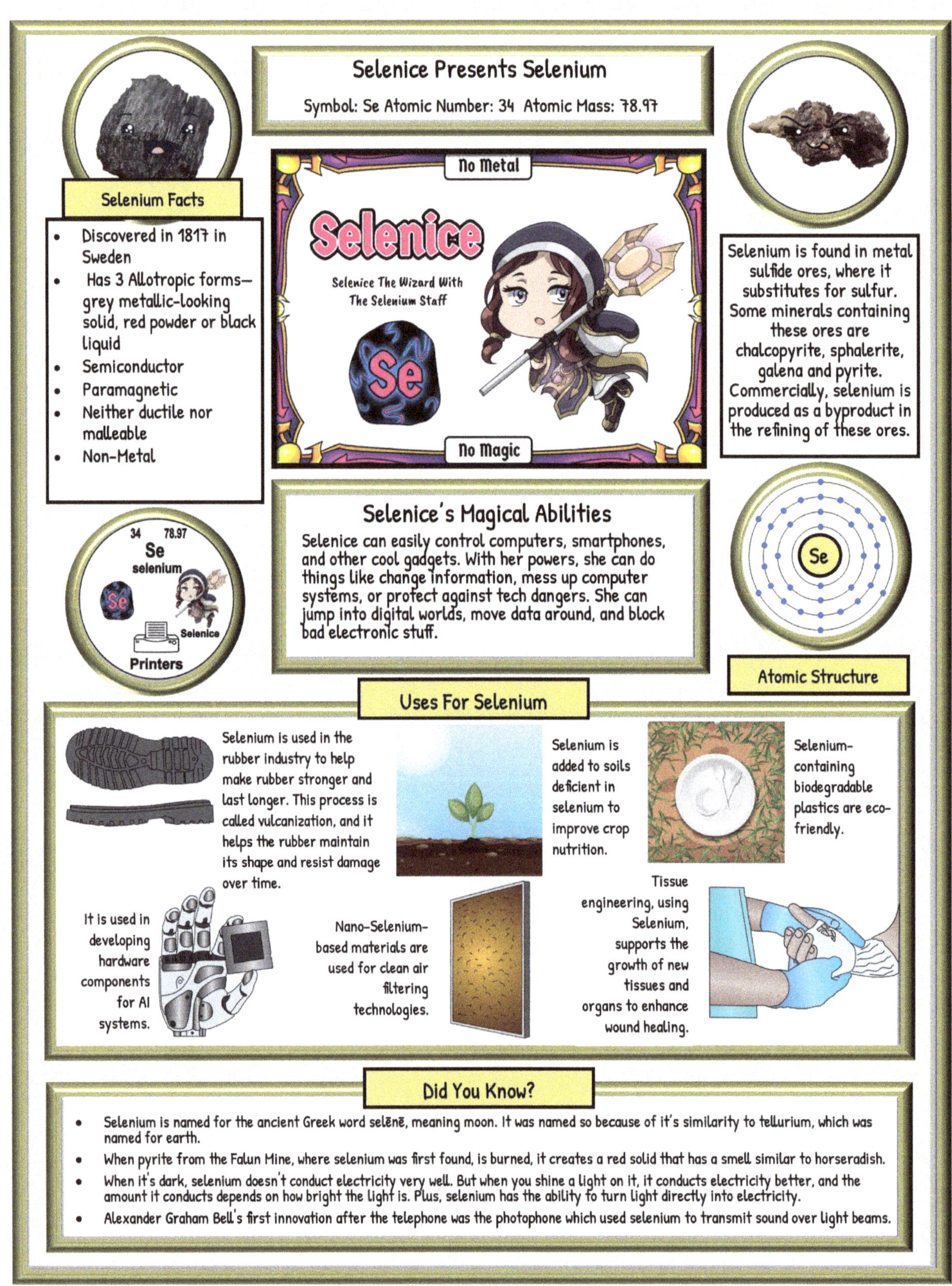
Selenice Presents Selenium

Symbol: Se Atomic Number: 34 Atomic Mass: 78.97

No Metal

Selenice

Selenice The Wizard With The Selenium Staff

Se

No Magic

Selenium Facts

- Discovered in 1817 in Sweden
- Has 3 Allotropic forms—grey metallic-looking solid, red powder or black liquid
- Semiconductor
- Paramagnetic
- Neither ductile nor malleable
- Non-Metal

Selenium is found in metal sulfide ores, where it substitutes for sulfur. Some minerals containing these ores are chalcopyrite, sphalerite, galena and pyrite. Commercially, selenium is produced as a byproduct in the refining of these ores.

34   78.97
Se
selenium
Se   Selenice
Printers

Selenice's Magical Abilities

Selenice can easily control computers, smartphones, and other cool gadgets. With her powers, she can do things like change information, mess up computer systems, or protect against tech dangers. She can jump into digital worlds, move data around, and block bad electronic stuff.

Atomic Structure

Uses For Selenium

Selenium is used in the rubber industry to help make rubber stronger and last longer. This process is called vulcanization, and it helps the rubber maintain its shape and resist damage over time.

It is used in developing hardware components for AI systems.

Nano-Selenium-based materials are used for clean air filtering technologies.

Selenium is added to soils deficient in selenium to improve crop nutrition.

Tissue engineering, using Selenium, supports the growth of new tissues and organs to enhance wound healing.

Selenium-containing biodegradable plastics are eco-friendly.

Did You Know?

- Selenium is named for the ancient Greek word selēnē, meaning moon. It was named so because of it's similarity to tellurium, which was named for earth.
- When pyrite from the Falun Mine, where selenium was first found, is burned, it creates a red solid that has a smell similar to horseradish.
- When it's dark, selenium doesn't conduct electricity very well. But when you shine a light on it, it conducts electricity better, and the amount it conducts depends on how bright the light is. Plus, selenium has the ability to turn light directly into electricity.
- Alexander Graham Bell's first innovation after the telephone was the photophone which used selenium to transmit sound over light beams.

MEET THE ALCHEMICAL WIZARDS

Arkyn-Arsenic

Aszrad-Astatine

Barsena-Barium

Berwyn-Beryllium

Cadmus-Cadmium

Crowmist-Chromium

Hallam-Hafnium

Nelushia-Neodymium

Paedin-Palladium

Polgarn-Polonium

Praethec-Praseodymium

Proctor-Promethium

Raele-Radium

Renkin-Rhenium

Samarinda-Samarium

Selenice-Selenium

Silonar-Silicon

Taltra-Tantalum

Tephen-Technetium

Terlin-Terbium

Thanelen-Thallium

WIZARDS WITH ELEMENTAL TIPPED STAFFS FROM BOOK 2

Magical Elements of The Periodic Table

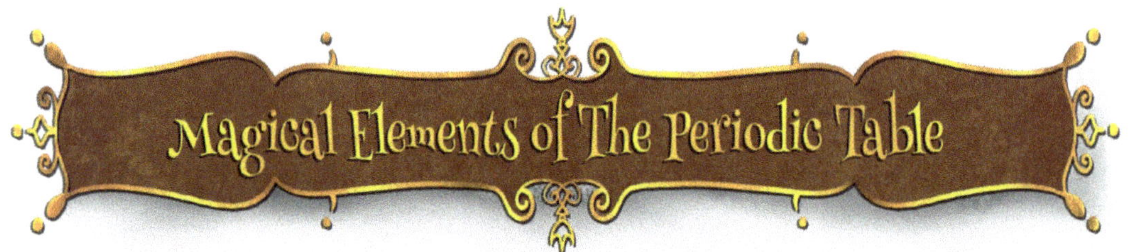

Create Your Own Magical Dragon Elemental

Selenice The Wizard With The Selenium Staff

Symbol: Se Atomic Number: 34 Atomic Mass: 78.97

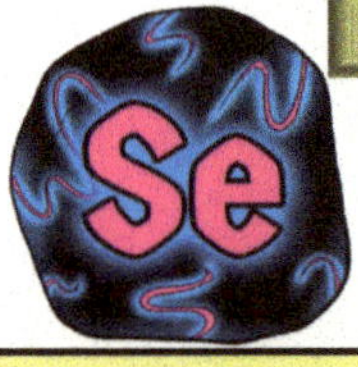

Magical Elemental Symbol

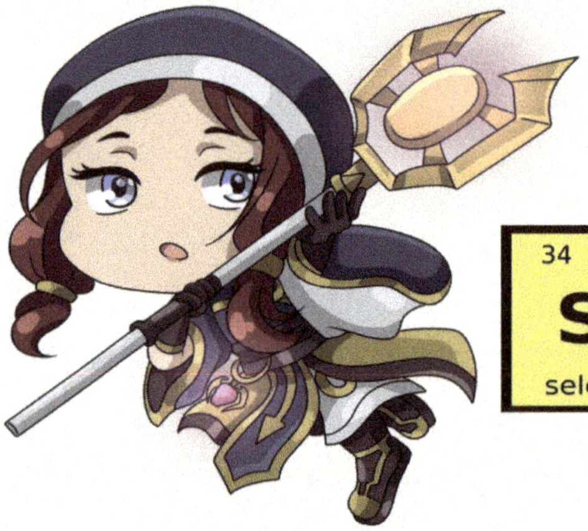

34 78.97
Se
selenium

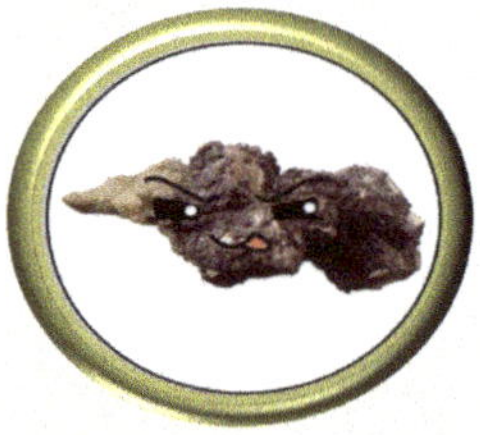

Selenium is a Non-Metal

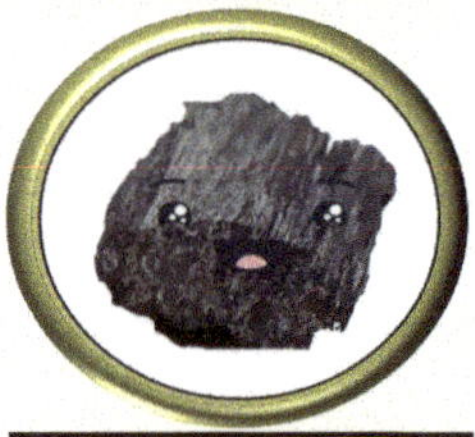

Extracted From Chalcopyrite

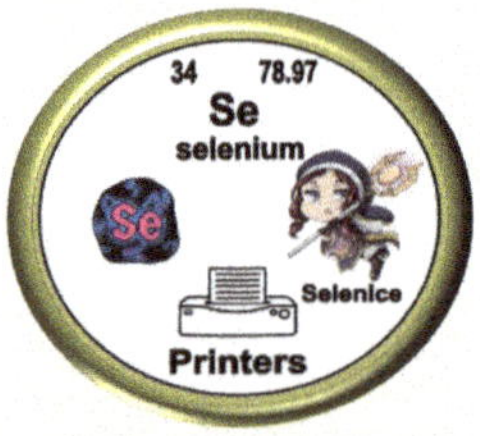

Selenium Periodic Symbol

Selenice's Magical Abilities

Selenice can easily control computers, smartphones, and other cool gadgets. With her powers, she can do things like change information, mess up computer systems, or protect against tech dangers. She can jump into digital worlds, move data around, and block bad electronic stuff.

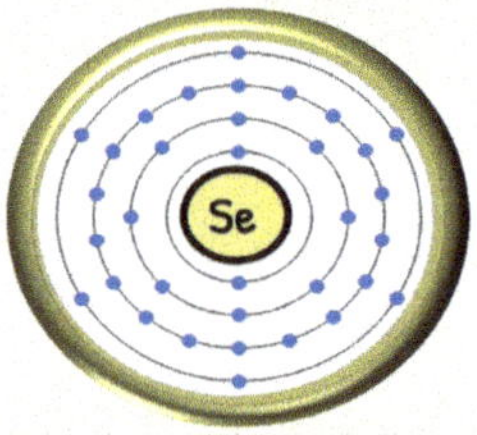

Atomic Structure

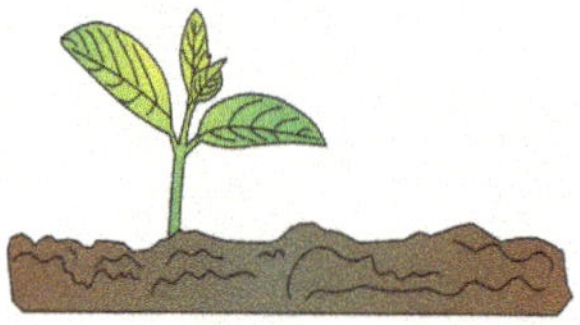

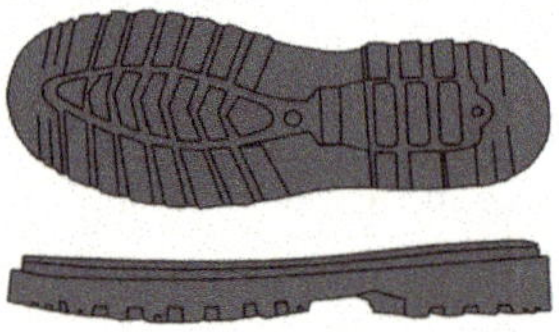

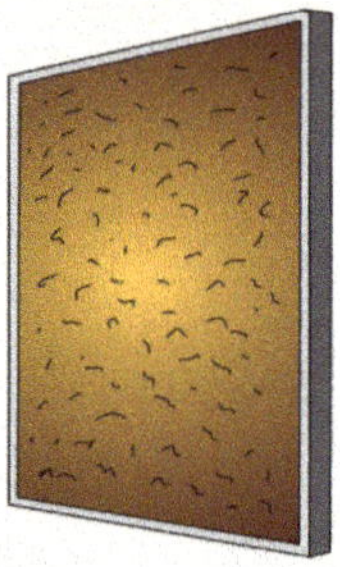

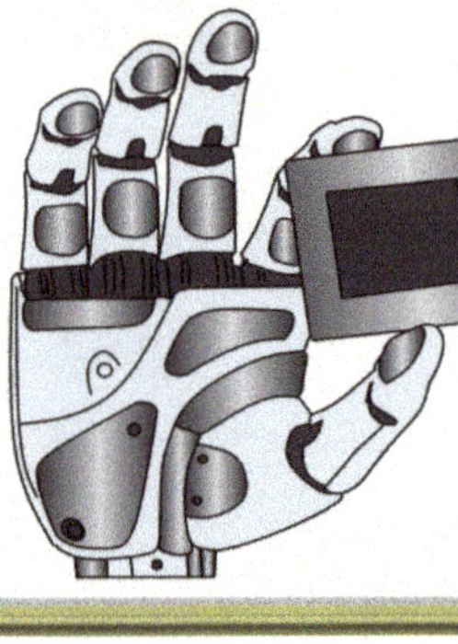

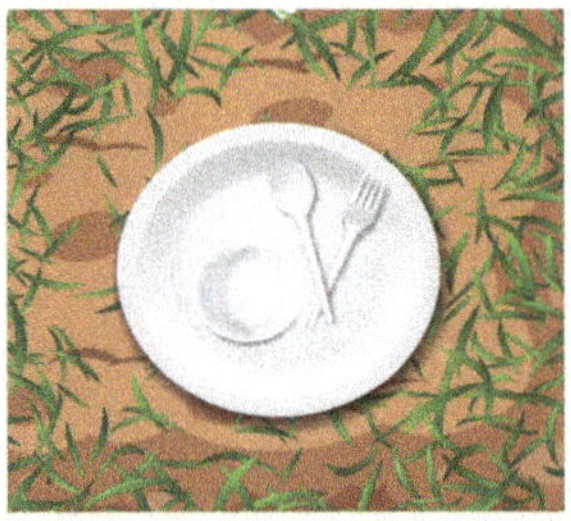

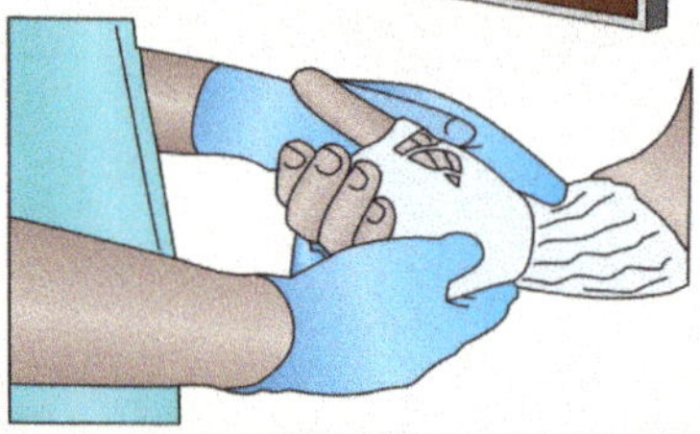

Students may either use a program like power point to cut and paste clip art into a Magical Wizard Elemental Blank or, if they wish, they may draw everything themselves.

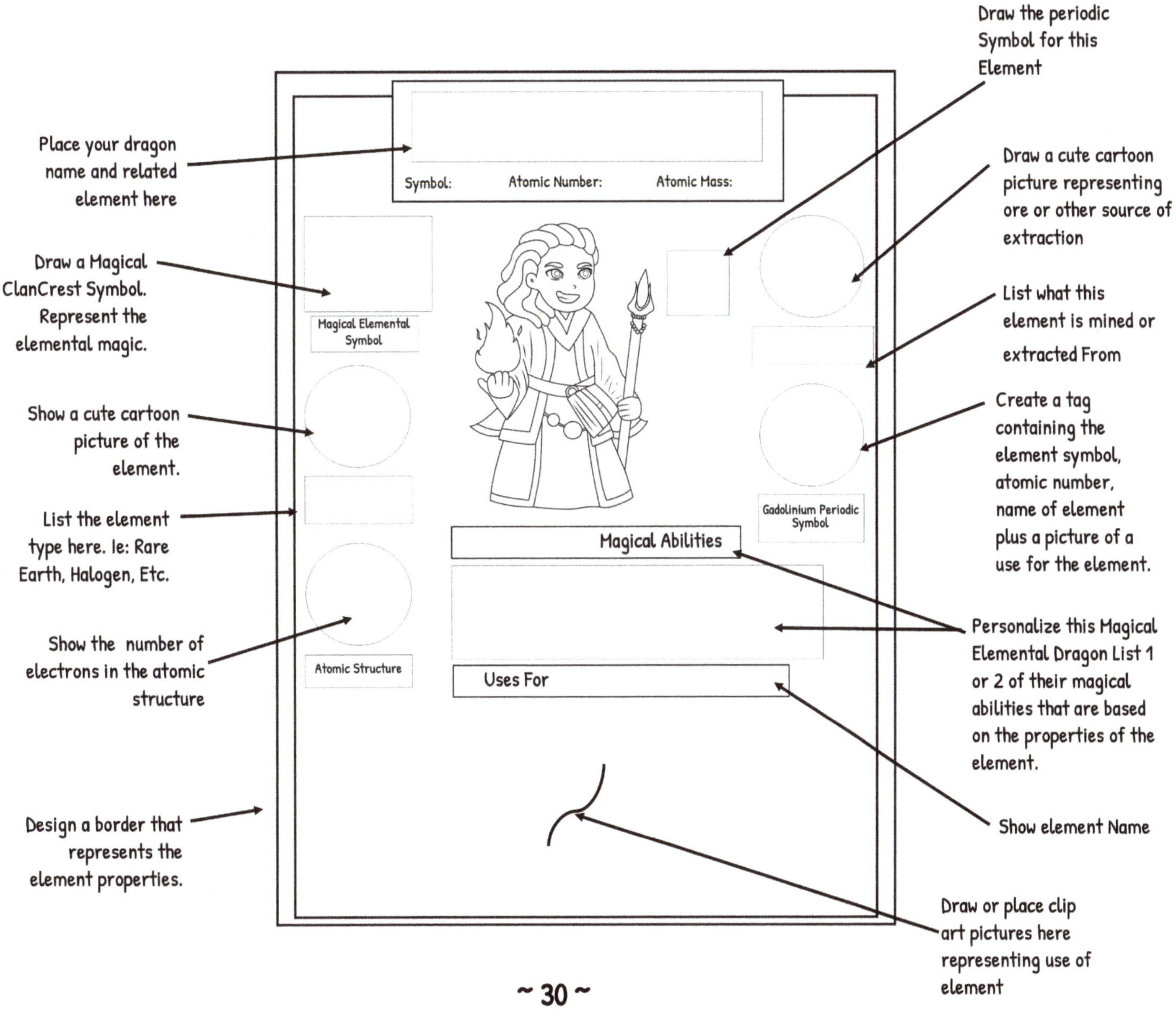

Draw the periodic Symbol for this Element

Place your dragon name and related element here

Draw a cute cartoon picture representing ore or other source of extraction

Draw a Magical ClanCrest Symbol. Represent the elemental magic.

List what this element is mined or extracted From

Show a cute cartoon picture of the element.

Create a tag containing the element symbol, atomic number, name of element plus a picture of a use for the element.

List the element type here. Ie: Rare Earth, Halogen, Etc.

Show the number of electrons in the atomic structure

Personalize this Magical Elemental Dragon List 1 or 2 of their magical abilities that are based on the properties of the element.

Design a border that represents the element properties.

Show element Name

Draw or place clip art pictures here representing use of element

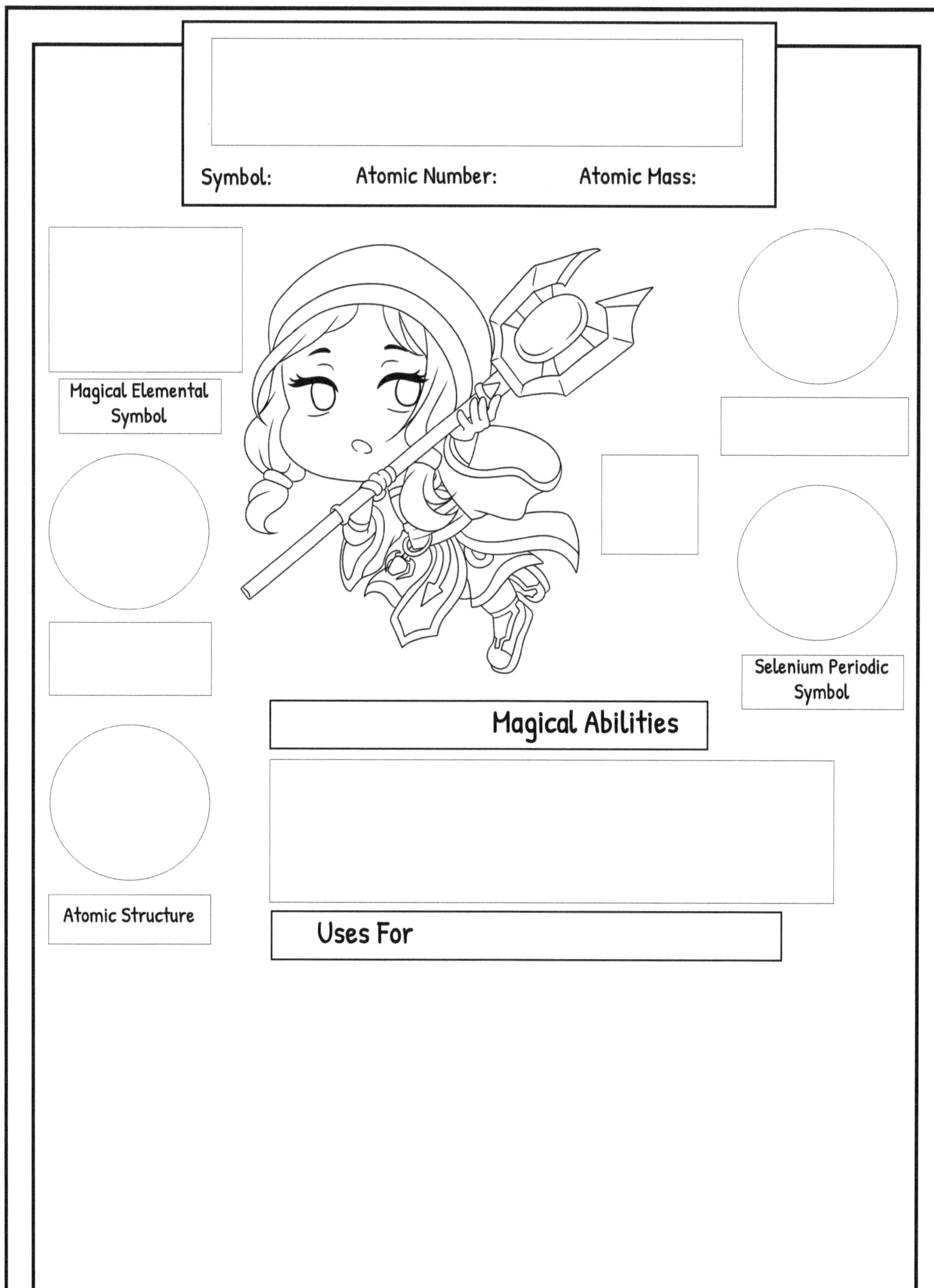

Symbol:
Atomic Number:
Atomic Mass:
Magical Elemental Symbol
Selenium Periodic Symbol
Magical Abilities
Atomic Structure
Uses For

Magical Wizard Elemental Research Sheet

Before starting your Magical Wizard Elemental graphics page, do some research on your chosen element.

Name of Magical Wizard:	
Wizards's Magic Power Based on the Element's Properties:	
Magical Elemental Symbol:	
Element Name:	
Element Symbol:	
Atomic Number:	
Atomic Mass:	
What year and where was this Element discovered?	
Who discovered this Element?	
Element Group:	
Element Period:	
Element Family Name:	
State of Element At Room Temperature:	
What is Element Mined or Extracted From?	
Is Element Magnetic?	
Does Element Conduct Electricity?	
Where is the Element commonly found in Nature?	
What is 1 alloy of the Element? How used?	
What is 1 compound of the Element? How used?	
Name the most common use for this Element:	
Name a little known use for this Element:	
Name one more use for this Element:	
Interesting and Fun Facts:	

Magical Unicorn Elemental Research Sheet

Before starting your Magical Unicorn Elemental graphics page, do some research on your chosen element.

Name of Magical Unicorn:	Ghel The Gold Horn Unicorn
Unicorn's Magic Power Based on the Element's Properties:	Ghel can see past, present and future. She is empathic and can sympathize with the feelings of other. They say she has a heart of gold.
Magical Herd Crest Symbol:	An open heart with a Celtic Trinity Knot.
Element Name:	Gold
Element Symbol:	Au— Comes from Aurum which is the Latin word for Gold.
Atomic Number:	79
Atomic Mass:	196.97
What year and where was this Element discovered?	Around 4,600 BCE in Bulgaria
Who discovered this Element?	Unknown
Element Group:	11 on Periodic TAble
Element Period:	6 on Periodic TAble
Element Family Name:	Gold is a Noble Transition Metal
State of Element At Room Temperature:	Solid
What is Element Mined or Extracted From?	Quartz Veins. It is also found in gravel in streams.
Is Element Magnetic?	It is Diamagnetic. It's only weakly magnetized when placed in a magnetic field.
Does Element Conduct Electricity?	Gold is a great electrical conductor used in printed circuitry of computers.
Where is the Element commonly found in Nature?	One of the largest deposits is found in the United States in Arkansas.
What is 1 alloy of the Element? How used?	White gold is an alloy of gold, palladium, nickel and zinc.
What is 1 compound of the Element? How used?	Gold Phosphide is a semiconductor used in high power, high frequency applications and in laser diodes.
Name the most common use for this Element:	Jewelry
Name a little known use for this Element:	Acupuncture needles
Name one more use for this Element:	Gold is used in airbags in cars.
Interesting and Fun Facts:	Gold was used in ancient Egypt to fill decayed teeth. Gold thread is incorporated in astronaut spacesuits to protect them from the heat of the sun.

Write a paragraph below to describe your magical wizard elemental. Based on the information obtained from research of your chosen element, how did you determine your wizard's name? What are your wizard's magic powers? What are their likes/dislikes, strengths/weaknesses, personality traits? What color are your wizards clothes and why did you pick that color? What is your wizard's Magical Elemental Symbol?

Magical Unicorn Elemental Sample Description

Ghel The Gold-Horned Unicorn

This magical unicorn has a golden horn and hooves that glow like the sun. Her hide is honey-gold and her flowing mane and tail are golden-blonde.

Ghel is a member of the Metal Horn Unicorn Tribe from Unimaise. Gold is linked to the heart chakra because it holds a warm energy that brings soothing vibrations to the body to aid in the healing process. Ghel's Magical Herd Crest symbol is an open heart with a Celtic Trinity knot which indicates that she can see past, present and future. She is empathic and can sympathize with the feelings of others.

Being empathic does not make Ghel a weakling. She is brave with strong opinions and is a true champion to those she loves. She is known as the unicorn with the "heart of gold". When she places her horn on the heart of another, she senses their future.

The name Ghel is an Indo-European word which means yellow. The word "gold" most likely has its origins in the word "Ghel".

Gold is known to possess spiritual powers that bring happiness, peace, stability and luck to those who wear it. Scientists say that all the gold in the world comes from the collision of neutron stars.

Though most other magical unicorn elementals get along well with the gold horn unicorn herd; Ghel, and others like her, must be very careful around the Quick Silver Herd - as gold dissolves in mercury.

Do Your Middle Graders Want To Know More From The Magical Elementals About The Periodic Table?

Get the accompanying books in print at all online book stores. Get the books and accompanying activities at MagicalPTElements.

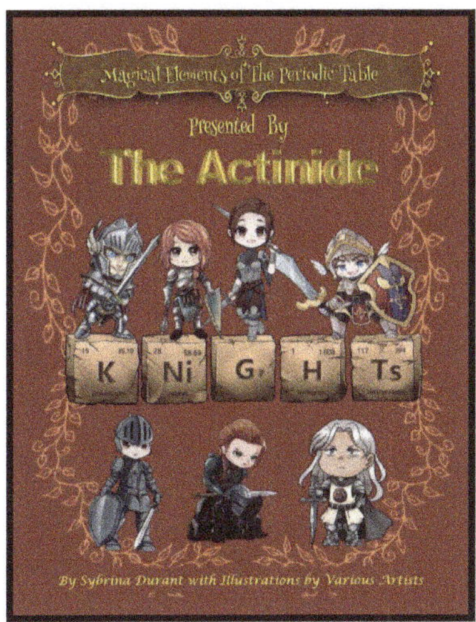

Get These Trading Cards Sets Featuring The Magical
Elementals Representing The Periodic Table Elements

at https://bit.ly/40oEUBr

Unicorns, Dragons, Wizards, Knights, or Goblins?

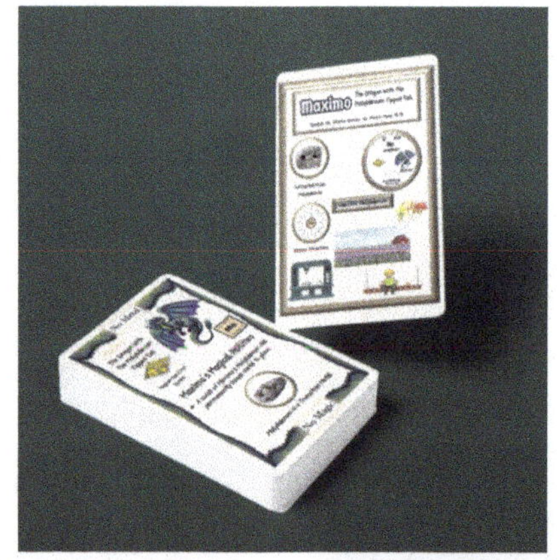

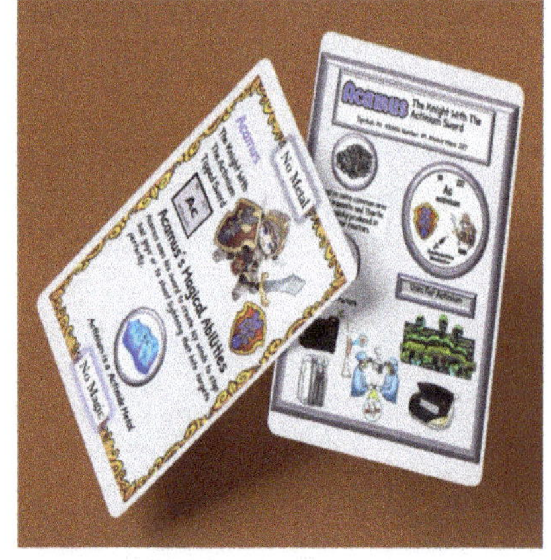

Collect Them All

These magical elementals are ready to help make learning the periodic table more fun. Get all books and related activities today.

Learn More About all of the Periodic Table Elementals. Get all of the "No Metal No Magic" Books Featuring Individual Elements at MagicalPTElements.com

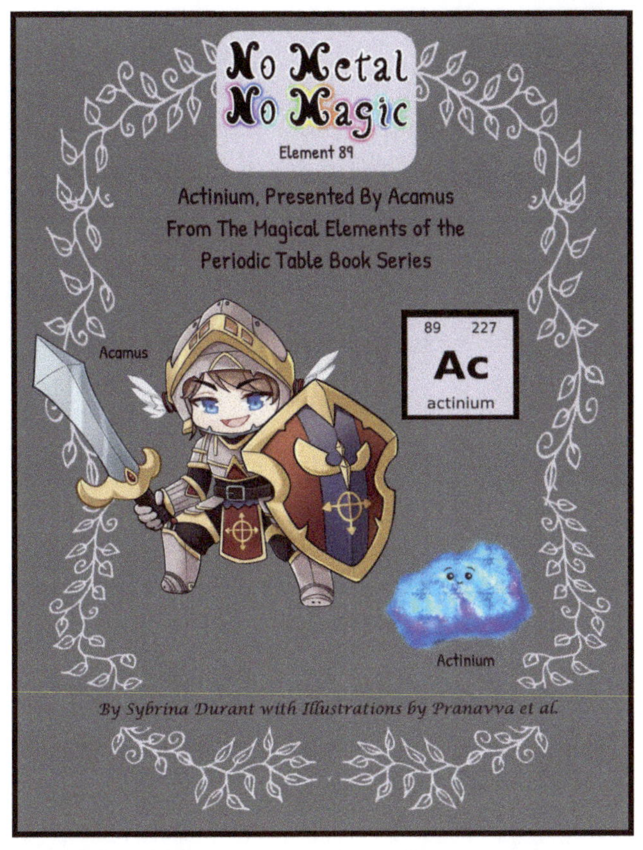

Also Available From Sybrina Publishing
Magical Elemental-Themed Periodic Table

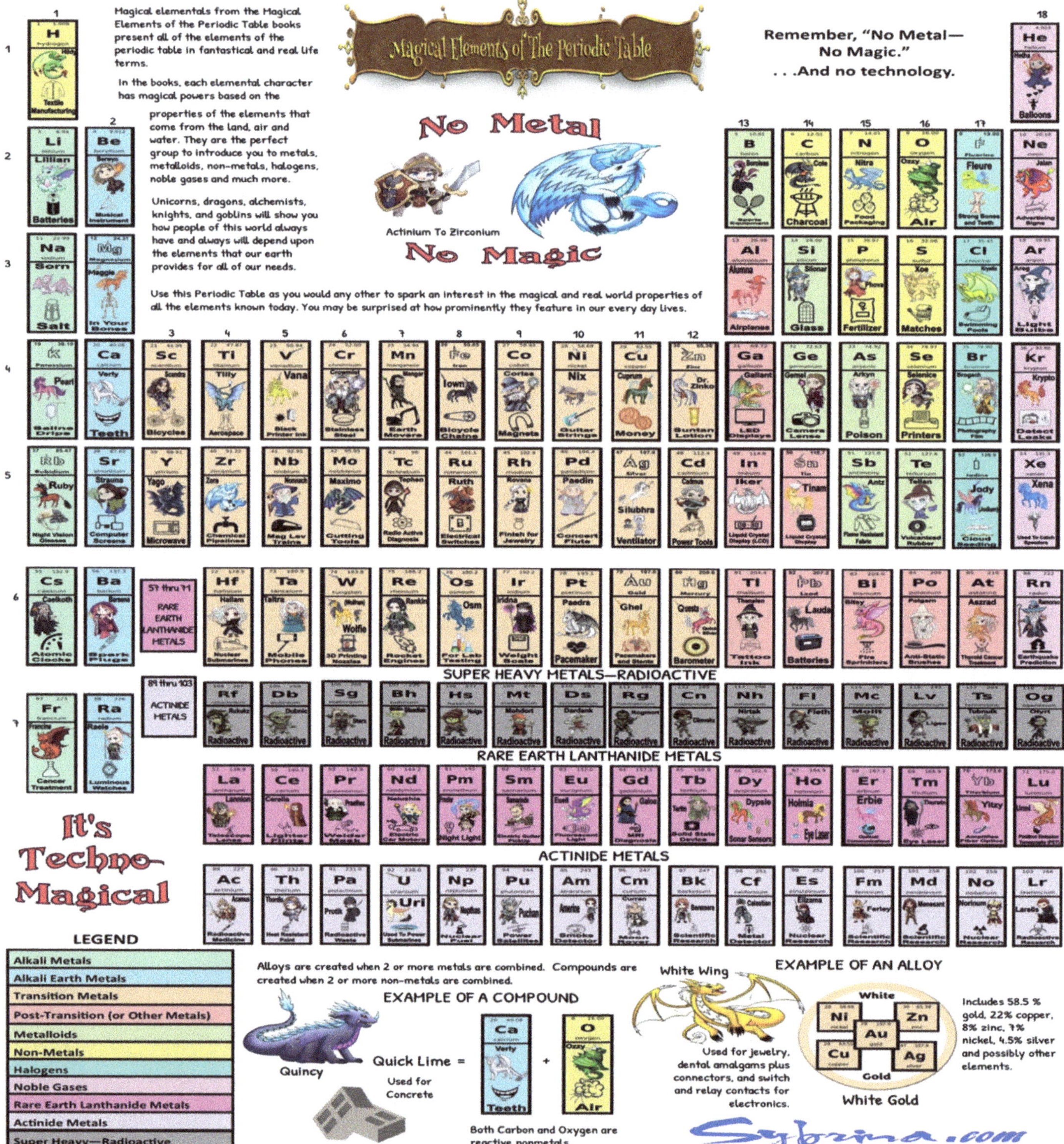

Would you like a 24" x 36" poster of the Magical Elemental-Themed Periodic Table from The Magical Elements of the Periodic Table books? The best place to get a high quality poster size print is at

https://bit.ly/49QMxBT It will look great on a classroom or kid's room wall.

Get These Fun Elemental Periodic Table Activities at

MagicalPTElements.

Unicorn Periodic Table Bingo—Comes with 32 unique Bingo cards. Magical Elementals Bingo comes with 36.

Magical Elemental Game Cards—Makes great prizes. Fun to trade, too.

1
2
3

plus

Unicorn Horn

Alphabet

CLIP ART FOR YOUR GRAPHICS

A
B
C

Also browse activities at
https://www.magicalPTelements.com
for all kinds of printable downloads to make learning fun.

Printable Magical Elemental Activity Downloads
Fun Way For Students To Learn The Elements Of The Periodic Table

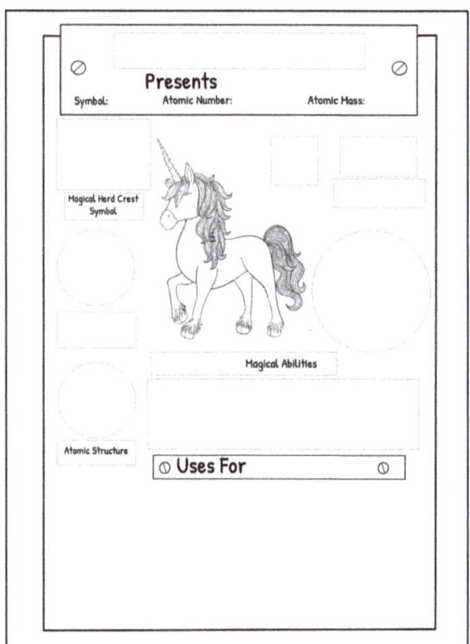

Blank Unicorn Element Card

Sample Unicorn Element Card

Blank Research Sheet

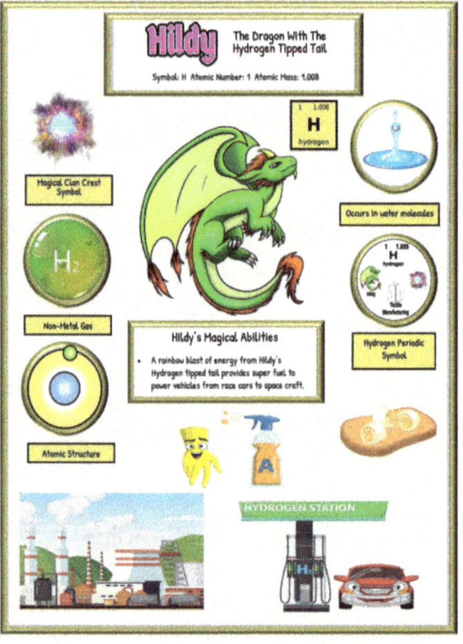

Sample Dragon Element Card

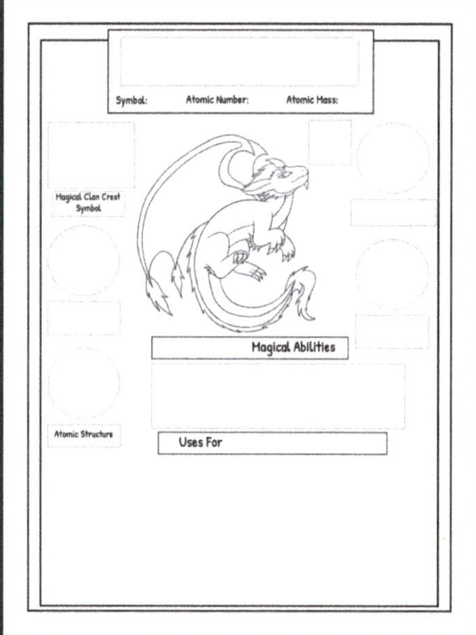

Blank Dragon Element Card

Blank Research Sheet

Using the sample Magical Elemental cards provided, have students select an element from the Periodic Table and a Magical Elemental Card Blank to create their own Magical Elemental Card. The blank and sample cards do not have to match.

You will receive a pdf containing either 26 unicorn or 26 dragon sample cards and blanks to be printed on 8 1/2 x 11 sized paper or card stock. The pdf also contains a Magical Elemental Research Sheet for the students to work on before creating their unique Periodic Table Elemental. They will also write a short paragraph describing their Unicorn or Dragon Elemental from that research.

Get These Fun Elemental Periodic Table Activity Sheets at *MagicalPTElements.com*

Don't Forget To Get A Tee Shirt

Featuring Your Favorite of the

118 Elements from the Periodic Table

Available in adult and kid sizes in many colors.

https://amzn.to/47NVZWN

This is the Selenice-Selenium Tee Shirt Graphic

Get it at https://www.amazon.com/dp/B0D9C4F9XQ

Dear Reader

I hope the "No Metal No Magic Element 2 — Hetha, from The Magical Elements of the Periodic Table Series Presents Helium" with illustrations by Pranavva et al, has helped you learn some fun and interesting things about the magic of the element, Helium.

This is one of what will eventually be 118 books featuring periodic table elements presented by unicorns, dragons, wizards, knights and goblins. Keep checking regularly. Every one of the elements are amazing and very necessary to our

Techno-magical.

A lot of research went into every page of this book as well as the Magical Elements of the Periodic Table Books. There are just too many references to publish in this book but you can read and research them all at MagicalPTElements.com/MAUPT or /MDAPT or /MW1PT or / MW2PT or /MAKPT. There, you can also access book related activity sheets and games to help make the learning process more fun.

Get ready made trading cards, lapel pins, tee shirts and more based on this book from Sybrina Publishing's No Metal No Magic Collection at Zazzle - **http://bit.ly/3km64Wg**

Would you like a 24" x 36" poster of the Elemental-Themed Periodic Table in this book? The best place to get it is at **https://bit.ly/49QMxBT** They have the sharpest images of any other poster printer around.

The Magical Elements of the Periodic Table books came into existence because of my Blue Unicorn—Journey To Osm books. If it weren't for their magical powers, based on the properties of the metals of their horns and hooves, I would have never come up with the idea to relate magical creatures to the periodic table. There's a metal horn unicorn story for every age group and they are all available at MagicalPTElements.com

If you enjoyed this book
please leave a nice review
at your favorite online book site.

No Metal No Magic

Song Lyrics

No metal, no Magic

No metal, no Magic

I can think of nothing more tragic

Than to have no metal or no magic

Metal makes everything magical.

Just ask a unicorn. . .

Preferably, one with a metal horn.

They'd say No metal, No magic.

Metal makes everything techno magical.

No metal, No magic

for two-leggers or unicorns.

No metal, No magic

Metal makes everything techno magical.

No metal, No magic

It's techno magical.

No metal, No magic

It might be very hard to believe but with

No metal, No magic

There'd be no technology.

No metal, No magic

Listen to this song at https://youtu.be/tcB8KDWAd8w

Watch the book trailer at https://youtu.be/NIX9fE7GJRI

Blue Unicorn

Ebooks, Audio and Print Books Available at all online book stores.

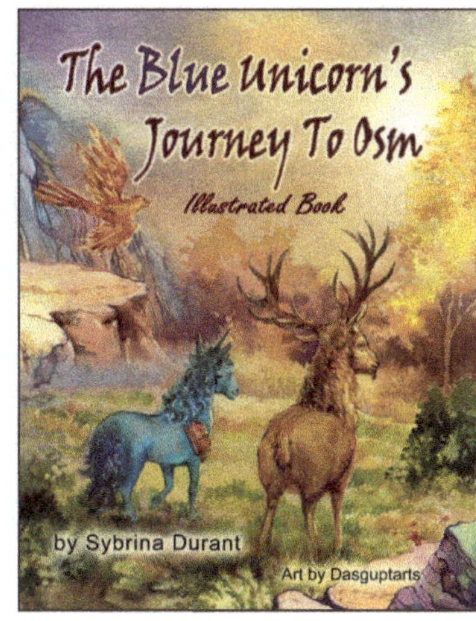

Illustrated
Book

Audio
Book

'Read & Color' Book
For Teens

Unicorn Periodic
Table Book

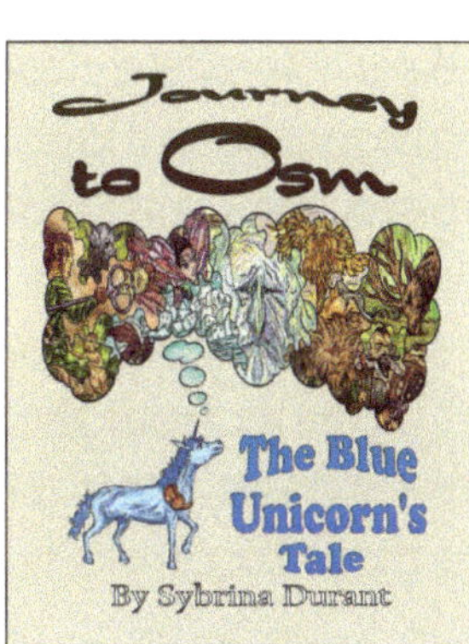

Fantasy
Novel

Coloring Book & Character
Introduction

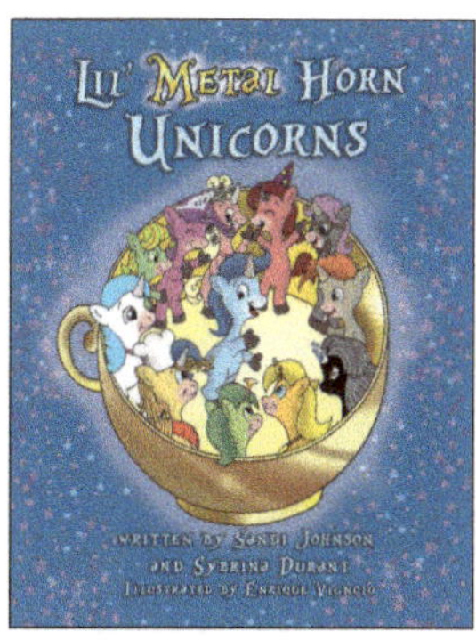

Picture Book For
Kids

Get these and more at
MagicalPTElements.com
and Sybrina.com

Journey To Osm—The Blue Unicorn's Tale

Back then, most places throughout MarBryn had wizards and sorcerers of some ability, or another. Most were trained in the ways of magical arts by the unicorns as part of their outreach program. Some two-leggers developed practical magical skills like making delicious feasts of tasty food appear out of thin air or purifying murky water around the land.

Others went through more extensive training to learn battle magic—like shooting powerful streams of energy from their swords.

Some were taught the art of holding the glow of the sun in magical globes, bringing light into the dark of night. These magical lights warded off the evil beings that were new and frightening products of dark magic.

With the rise of the sorcerer Magh, magical defense arts had become more important. The highest level of magical training involved sensing when others were in danger and learning to see into the future. Very few two-leggers ever reached that level because magic wasn't inherent in them the way it was for the unicorns. The metal of their horns and hooves were part of them as well as the very makeup of their blood, but two-leggers relied on learned magic via potions, charms, and incantations that required help from ingredients and forces more mystical in nature than any two-legger was ever born to be. Of course, controlling magic and projecting your intentions went far beyond merely following a recipe of sorts. It took being in touch with nature and the various elements to get the response a wizard desired. Much trial and error went into it, as well as faith and trust and the motives of the spell caster.

Magic was and is a practice that is never quite perfected even for the unicorns who must continue to hone and learn how to harness their powers. On rare occasion a wizard and unicorn had formed enough trust and a steadfast bond that prompted the unicorn to gift the wizard with a wand or staff embedded with the smallest sliver of metal from one of their hooves. This was rare but had happened and of course so had the desire for more power. Magh wasn't the first sorcerer with lust for more and throughout history there had been a handful of heinous acts against unicorns from those seeking their magic. Prior to Magh those wizards had failed to circumvent the protections nature had infused unicorn magic with so even after harvesting metal from their horns or hooves these sorcerers had gone mad trying to bend the will of nature and actually use their ill-gotten gains.

Magh, too, had gone mad or perhaps he'd already been so, but somehow he'd managed to harness the metal he harvested from his victims and continued to grow stronger rather than completely lose his mind like the others. How exactly remained a mystery to the unicorns and everyone else.

The wizards who'd honed their craft with the help and blessing of the unicorns tried to solve the riddle and even the best of them failed to uncover that secret. Some of MarBryn's natives took to magic naturally, while others struggled with the concept but no matter how adept they were. All of them fought valiantly against Magh's magic because with even a shred of knowledge they understood how the power shift would ultimately play out. They fought to the end but, in the end, only one sorcerer remained in MarBryn and now, Magh was in total control. But still, he was not satisfied. He wanted to control every living creature in the land.

Get The Novel at MagicalPTElements.com